Scientific Visualization
and Graphics Simulation

Scientific Visualization and Graphics Simulation

Edited by
Daniel Thalmann

*Swiss Federal Institute of Technology,
Lausanne, Switzerland*

JOHN WILEY & SONS
Chichester · New York · Brisbane · Toronto · Singapore

Other Wiley Editorial Offices

John Wiley & Sons, Inc., 605 Third Avenue,
New York, NY 10158-0012, USA

Jacaranda Wiley Ltd, G.P.O. Box 859, Brisbane,
Queensland 4001, Australia

John Wiley & Sons (Canada) Ltd, 22 Worcester Road,
Rexdale, Ontario M9W 1L1, Canada

John Wiley & Sons (SEA) Pte Ltd, 37 Jalan Pemimpin 05-04,
Block B, Union Industrial Building, Singapore 2057

Library of Congress Cataloging-in-Publication Data:

Scientific visualization and graphics simulation / edited by Daniel Thalmann.
    p. cm.
    Includes bibliographical references.
    ISBN 0 471 92742 2
    1. Science—Methodology.   2. Visualization.   3. Digital computer simulation.
    4. Computer graphics.   I. Thalmann, Daniel.
    Q175.S4243   1990
    501′.5118—dc20   90-12445
                                                                     CIP

British Library Cataloguing in Publication Data:

Scientific visualization and graphics simulation.
    1. Computer systems.   Graphic displays
    I. Thalmann, Daniel
    006.6

    ISBN 0 471 92742 2

Printed in Great Britain by Biddles Ltd, Guildford

Contents

15   Graphics Visualization and Artificial Vision

16   Collaboration Between Computer Graphics and Computer Vision

17   Graphics Simulation in Robotics

18   Characteristics of a Model for the Representation of a Production Installation

Preface

Scientific Visualization is a new approach in the area of numerical simulation. It allows researchers to observe the results of numerical simulations using complex graphical representations. Visualization provides a method for seeing what it is not normally visible, e.g. torsion forces inside a body, wind against a wall, heat conduction, flows, plasmas, earthquake mechanisms, and molecules. The purpose of this book is to present the techniques of computer graphics, image synthesis and computer animation necessary for visualizing phenomena and mechanisms in sciences and medicine. The book provides engineers, scientists, physicians, and architects with fundamental concepts allowing them to visualize their results using graphics and animation. Complete examples of visualization are presented in biology, chemistry, physics, medicine, and robotics.

This book is the first book dedicated to scientific visualization with an emphasis on basic geometric, animation and rendering techniques specific to visualization as well as concrete applications in sciences and medicine. The book is based on special lectures on Scientific Visualization and Graphics Simulation organized at the Computer Graphics Laboratory of the Swiss Federal Institute of Technology in Lausanne. Each chapter has been written by one or several experts in a particular aspect of visualization.

The books starts with an overview of graphics workstations and processors. Then, several fundamental problems of computational geometry are presented. The third chapter tries to answer the question "When does a surface look good ?". As volumes are essential in scientific visualization, three chapters are dedicated to various aspects related to representing volumes: solid modelling, finite element theory and volume rendering techniques. Then, special methods for modelling natural objects are described such as fractals, particle systems, and modular maps. Basic and advanced techniques in Computer Animation are then described including keyframe animation, algorithmic animation and human animation. The next chapter presents robotics methods for task-level and behavioral animation. Then, the Astrid system for graphical representation of numerical simulations is described in a specific chapter. Additional chapters are dedicated to typical applications of visualization and graphics simulation: visualization of flow simulations, visualization and manipulation of medical images, visualization of botanical structures, and visualization of complex molecular models. Two chapters are dedicated to computer vision and its relationship to computer graphics. Finally, the last two chapters presents the graphics simulation of robots and production installations.

The editor would like to thank the director of the Swiss Federal Institute of Technology for his support in organizing the lectures which led to the publication of this unique book.

Daniel Thalmann

1

Architecture of Graphic Systems

J.D. Nicoud
Ecole Polytechnique Fédérale de Lausanne

1.1 Introduction

Although a primitive graphic CRT display was used on the 50´s Whirlwind computer built at MIT, Ivan Sutherland was the first in 1963 to analyse interactive displays correctly and propose data structures and interaction techniques (Sutherland 63). The first system using these concepts was only implemented in 1968 at Harvard University.

The next major step was the Alto station by Xerox Parc (Thacker 82), which was developed from 1974. It included dedicated microprograms for bit bloc transfers (bit–blt) and developed most of the concepts associated with bit–mapped graphics. Present personal computers offer the same performance, and are adequate for text processing, 2–D drawings and 3–D wire objects representation.

Large scale integrated circuits have opened up a new era of highly sophisticated 3–D stations. Jim Clark was the first to build a dedicated Geometry Engine integrated circuit (Clark 82); now more than a million transistors can be found in a microprocessor like the i860, which includes a few efficient graphic primitives (Grimes 89).

Functionally, an image or drawing must be stored as an abstract, device–independent form. Models use world coordinates and when a screen representation is required, the view point, the viewing angle, the light sources as well as the window must be specified and a set of transformations take place before the flattened representation is clipped into the bit mapped screen (figure 1.1).

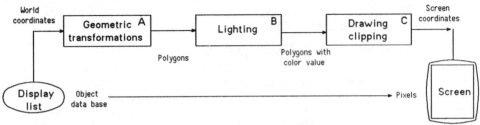

Figure 1.1. Transformation steps for a 3-D object

The Geometry Engine (Clark 82) is a neat solution for bloc A; a pipeline of a dozen chips including each four floating point units does the geometric transformation and the clipping.

The graphic pipeline transforms a hierarchical description of a scene into a device–dependant particular view. It consists of the following steps :

- traversing of the graphic tree with the instanciation of objects at their actual positions
- converting vertices into universal coordinates
- clipping polygons according to the view volume
- calculating the illumination in each vertex
- decomposing the polygons into horizontal trapezoids
- filling these trapezoids with color interpolation (Gouraud shading)
- painting pixels into the frame buffer (rasterization) with taking care of multiwindow and visibility

Due to high bandwidth requirements, hardware is required for the last stages of mapping on the screen whereas software is used for the complex transformations. However, because of the increasing need for performance and the decreasing cost of hardware, the boundary is moving and modern high performance graphic workstations include a compact, but complex and sophisticated amount of hardware.

The aim of this chapter is to present the major hardware solutions for computing and displaying graphic information, limiting ourselves to the parts close to the screen and not mentioning the microprocessor unit (MPU) hardware and software environment (figure 1.2).

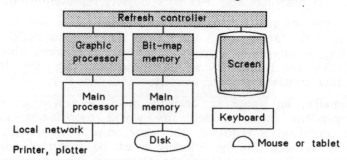

Figure 1.2. Block diagram of a graphic workstation

1.2 Screen technologies

The success of bit–mapped graphic displays is due to the low cost of CRT screens. An electron beam is focused on a phosphor–coated screen. A deflection system makes the spot move randomly or over scanned lines. Scanning the same image again and again is necessary in order to refresh the non–persistent phosphor. A rate greater than 50 Hz (70 Hz for black on white screens) avoids flickering. Long persistent phosphor is used sometimes to lower the refresh rates, but this spoils the interactive activities.

Random scan CRTs became successful in the days when memory was slow and expensive. Dot coordinates are sent to interpolators controlling digital–analogue (D/A) converters in order to draw segments with a flickering rate which depends on the number of segments to be displayed (figure 1.3). Filled areas, colors as well as shading are possible with restrictions and only if special solutions are used.

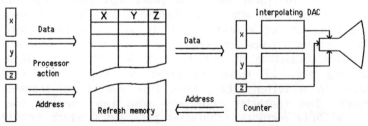

Figure 1.3. Random scan CRT

Raster scan CRTs display objects as a set of dots across each scan line (figure 1.4). Resolution depends on the number of scan lines and modulation frequency.

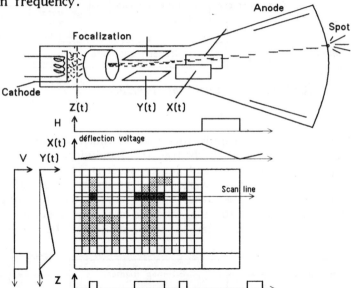

Figure 1.4. Raster scan CRT

Designing the interface is easy once the problem of serializing a copy of the screen in memory has been solved. This was expensive when memory chips only had a small capacity, but nowadays this part of the interface is very cheap (see next section). The screen, which generates the high voltage ramps for deflecting the spot (figure 1.4), must be provided with two synchronizing pulses H and V.

Flat Plasma screens or Liquid Crystal Displays (LCD) are not yet competitive for large size screens. Light emitting, light reflecting or light absorbing dots are formed at the intersection of orthogonal lines. Large quantities of electronic devices are required for driving, selecting and scanning these dots. Manufacturers usually embed an interface which makes the screen look like an addressable memory or like a CRT screen with H and V synchronization pulses.

1.3 Screen interfaces

1.3.1 Alphanumeric screens

Alphanumeric interfaces built 10 years ago had a memory for the text on the screen which was down-loaded from the main memory of a central computer running a text editor. In synchronization with the spot scanning the screen, lines of text were transferred towards a small buffer typically consisting of two lines of 80 characters. While one line was circulating e.g. 12 times in order to display the vertical dots corresponding to a letter, the next line was being transferred from page memory. The character code and scan line number was used to address a read only memory (character generator) which provided a fast shift register with the line of dots to be displayed. The high frequency part hence only concerned few circuits (figure 1.5).

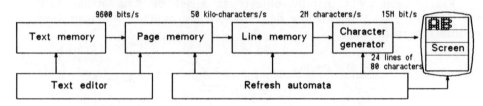

Figure 1.5. Alphanumeric terminal interface

1.3.2 Graphic screens

The architecture of a graphic interface is much simpler, but needs a large bit-map memory and a high transfer bandwidth, increased by the fact that graphic applications need larger screens. The bit-map memory is dual ported. The processor writes and reads on one side while the refresh automata reads on the other side (figure 1.6). A sequencer must arbitrate the simultaneous requests, and be sure the display will receive the information on time, in order to avoid snow on the screen.

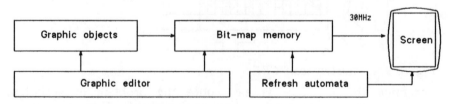

Figure 1.6. Graphic screen interface

The interface for a simple monochrome graphic display as found in most personal computers can be built with four 64k by 4 memory chips (figure 1.7). The display controller regularly provides the increasing screen addresses and generates a read transfer from memory into a shift register. The processor must frequently wait for the end of a word transfer to the screen, which reduces its efficiency.

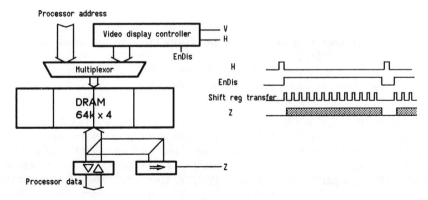

Figure 1.7. Bit-map graphic display

The width of the memory words defines the frequency of the transfers into the shift register used for serializing dots. For instance, if the screen is 640x480 and the memory words are 16 bits, one has the figures :

Vertical retrace frequency	60Hz	low flicker
Horizontal retrace frequency	30 kHz	60Hz × (480+20)
Pixel frequency	24 MHz	30kHz × (640+160)
Shift register load frequency	1.5 MHz	24MHz / 16)
Memory transfer cycle	250ns	
Memory occupancy (μP slow–down)	30%	

For increased performance, interleaving of memory planes and special dynamic memory cycle (page mode, nibble mode) were used in the early 80´s. Videorams (Nicoud 87) provided an elegant solution which strongly contributed both to the success of graphic displays and to lowering their cost. The principle of a videoram is to include a long shift register (SAM), loaded by a special transfer cycle. This register is long enough to be loaded at each horizontal retrace only (figure 1.8). Hence, there is no tight synchronization problem, since the screen is blanked while the transfer occurs. Another noticeable benefit is that 95% of the memory is available for the processor.

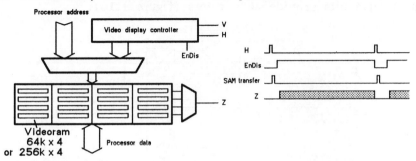

Figure 1.8. Bit-map graphic display using 4 videorams

Available Videorams have a size of 64kx4 and 256kx4. The internal register size is respectively 256x4 or 512x4.

1.4 Gray scale and color screens

Several bit–map memories can be connected in parallel to encode a dot intensity. A digital to analog (D/A) converter will provide gray scale levels to the screen (figure 1.9a). That simple organization is named planar but it has a drawback: in order to modify one pixel, several memory accesses are necessary.

A preferable solution is to group the screen memory into packets or "chunks" of conceculive bits, each representing a pixel (figure 1.9b). The interface of figure 1.7 must use four 4–bit shift registers with adequate wiring. The videoram interface (figure 1.8) is easier to modify : the output multiplexer is replaced by the D/A converter. Memory capacity and output bandwidth is increased four–fold. This may imply interleaving two or four memory planes.

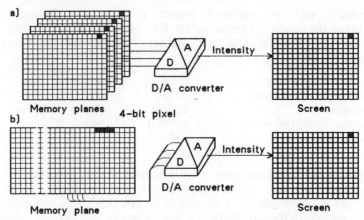

Figure 1.9. Black and white screen with a) 16 gray levels with planar organization (a) and packet (chunky) organization (b)

Color screens just need more bits. Shadow mask CRTs have three electron guns corresponding to the fundamental red–green–blue (RGB) colors. Twelve bits provide 4096 colors (figure 1.10).

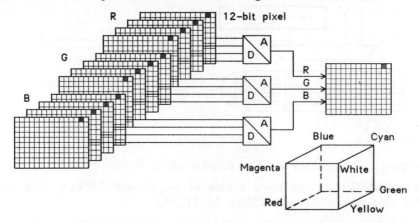

Figure 1.10. Direct 4096 color interface (planar organization)

For saving memory, and using more flexible logical color numbers, a color table is used in most cases. A fast RAM selects a limited number of colors among a large palette (figure 1.11). The same figure shows that packed organization is more frequently used for processing individual pixels at a faster rate.

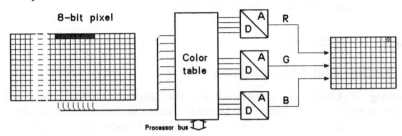

Figure 1.11. Color table for 256 colors among 4096 possible colors (packed or chunky organization)

Let us note that printers use a different set of primary colors: cyan, yellow and magenta (CYM), since colors are seen by reflected light, which is a subtractor process. The hue–saturation–value (HSV) model is frequently used for modelling, since it is easier to handle for operators (Hearn 86).

1.4.1   Antialiasing

Gray levels can be used in order to increase the apparent resolution of a screen and reduce the unpleasant "jaggies" of a line with a slow slope (figure 1.12).

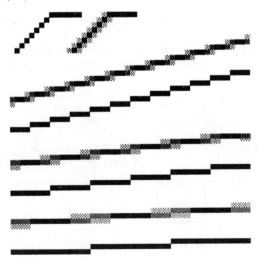

Figure 1.12. Antialiasing technique (to be also viewed from a certain distance)

This also allows for compensation of the line–intensity problem : a line at 45° looks thinner than horizontal or vertical lines. Another antialiasing solution proposed by Megatek uses a special screen with additional deflecting coils which allows for a fine correction of the deflection of the beam. This has the same effect as a higher resolution CRT, but the bandwidth is kept the same.

1.5   Graphic operations

1.5.1   Raster transformations

Moving a block of pixels, e.g. the mouse cursor, is a frequent operation which is the first to benefit from dedicated hardware. A bloc is not aligned on memory word boundaries (figure 1.13). Copying it to any coordinates requires masking and shifting over word boundaries : these operations are performed slowly on standard microprocessors.

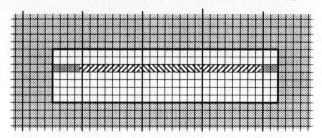

Figure 1.13. Decomposition of a bit bloc into memory words and part of words

Logical operations (AND, OR, NOT) frequently have to be executed at the same time. The "raster–op" operator performs a simple copy while carrying out the logical operation (figure 1.14a). The "BitBlt" operator is more general and includes clipping (Ingals 81) (figure 1.14b).

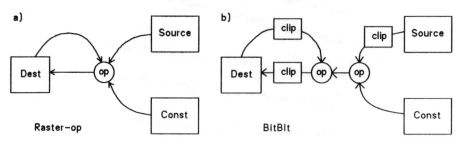

Figure 1.14. Raster operation (a) and BitBlt operation (b)

1.5.2   1 and 2-D objects

Drawing lines and circles, as performed efficiently by the Bresenham algorithm (Schweizer 87) can be implemented by a small amount of dedicated hardware (registers, counters) controlled by a microprogrammed unit. Splines and Bezier curves are handled by software, since they are not so frequent and would imply a lot of hardware.

1.5.3   3-D objects

Volumes represented as wire frames are old-fashioned. Surfaces are decomposed into triangles or polygons. Continuity conditions on the edges will define if edges must be visible or smoothened as well as possible.

1.5.4   Visibility

Clipping avoid to visualize parts of 2-D or 3-D objects which fall outside of the visualization window. Usually, software defines the coordinates of intersections of lines and boundaries. Hardware comparators can also inhibit the drawing when dots are outside. In the first case, time is lost computing intermediate dots. In the second, unuseful dots are drawn.

Hidden surface removal is an important operation for the visibility of 3-D objects. The processing is very simple if a large enough memory, the Z-buffer, stores for each pixel the distance toward the observer. For each dot of a drawn polygon, the distance is compared to the distance of the previous dot. If lower, the dot is not updated. The Z-buffer stores 16-bit integers or 32-bit floating point numbers when more precision is required.

1.5.5   Rendering

Shading algorithms interpolate between the colors of the vertices of the triangles or trapezes in which surfaces are decomposed, in order to render the smooth aspect of object illumination. Dedicated hardware does accelerate it. Ray tracing is a highly computing intensive operation in which rays are sent from the operator's eye and are reflected and absorbed according to the orientation and propriety of the encountered polygons. Multiprocessors are better suited for that application.

1.6   Graphic controllers and accelerators

Graphic controllers have simplified the building of graphic displays since 1980. They interface to the processor as programmable interfaces do. Control sequences trigger the drawing of lines, the filling of areas and the moving of data blocs. The most successful devices have been the EFCIS 9367, Nec 7220, Hitachi 68484, Intel 82786 and NS 8500 Graphic chip set. The drawback of these circuits is their slow action on groups of bits. A recent model, the Hitachi 64400, includes a direct path toward the processor and its main memory (figure 1.15).

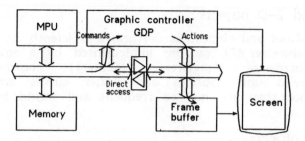

Figure 1.15. Typical display controller (Hitachi GDP 64400)

Improved drawing speed implies parallelism (Fuchs 87). The Pixel plane prototype uses hundreds of dedicated circuits, each handling a part of the bit–map memory (Poulton 85). All these circuits handle the functionality of bloc C in figure 1. Dedicated multiprocessors handling ray tracing have been proposed for bloc B. The Geometry Engine (Clark 82) is a neat solution for bloc A; a pipeline of a dozen chips including each four floating point units does the geometric transformation and the clipping.

1.7 Graphic processors

Instead of multiplying the chips, an alternative is to develop graphic processors which have efficient primitives for performing the operations of the graphic pipeline of figure 1. These processors were first derived from signal processor architecture (Texas TMS 34020) and included fast integer operations and raster–op primitives, suitable for 2–D graphics. The new trend for 3–D graphics is to use a general purpose RISC processor with fast floating point operations and dedicated instructions for drawing, Z–buffer handling and shading. The recently announced i860 (Grimes 89) (Case 89) is quite impressive in this respect, with its 128–bit floating point bus, 64–bit data transfer bus and parallel pipelined multiply and add operations (figure 1.16).

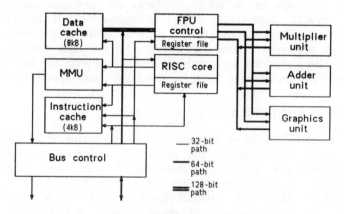

Figure 1.16. Block diagram of the Intel i860

1.8 High performance workstations

The first good graphic workstations in 1983 were built around a good microprocessor like the Motorola M68000. The next step was to add graphic accelerators for faster geometric transformations and quicker bit bloc transfers and operations. Dedicated hardware including many gate arrays and VLSI circuits was designed by the major manufacturers for their product line. The present leader is Silicon Graphics (Ackley 89), using both a Geometry engine consisting of a pipeline of floating point units, and a Raster engine which controls a videoram frame buffer of up to 56 bits per pixel. Multimode graphic processors handle color modes, formats and windows (figure 1.17).

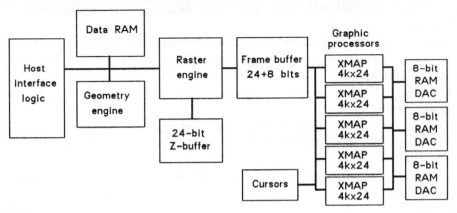

Figure 1.17. Iris GT block diagram

Another approach proposed by Ardent and others is to combine an array processor and a raster graphic controller. The user reaps the benefits of the array processors both for simulation and its graphic representation. Several future workstations will include high performance processors like the i860. A next step may be to use a large number of powerful graphic processors sharing fast serial communications, the functions being adequately distributed among the processors (Schweizer 88).

1.9 Conclusion

Graphic stations are still in infancy. On 2-D graphics, users of schematics editors, VLSI layout and mechanical drawing are limited by screen resolution. Progress in this respect will be slow. 3-D graphics are improving rapidly, but the requirements for realistic scene and technical objects representation is very high, especially in the case of animation. Many competitive circuits and stations will appear in the future, and there will be portable low cost, flat screen, 1024 by 1024 colored 3-D visualization stations before the end of the century.

References

K. Ackley, "The Silicon Graphics 4D/240GTX Superworkstation" IEEE Computer Graphics and Application, July 1989, pp.71–83

B. Case, "Evaluating the i860 Graphics Unit", Microprocessor Report, Vol 3 No 4, April 1989, pp 14–17

J. Clark, "The Geometry Engine: A VLSI Geometry system for Çraphics", Computer Graphics, Vol 16 No 3, 1982, pp.127–133

H. Fuchs "VLSI for Graphics", in *Techniques for Computer Graphics*, D.E. Rogers et al. editors, Springer Verlag 1987, pp281–294

J. Grimes, "The Intel i860 64–bit Processor: A general–purpose CPU with 3D Graphic Capabilities" IEEE Computer Graphics and Application, July 1989, pp.85–94

D. Hearn, P.Baker, "Computer Graphics", Prentice Hall, 1986, 351p.

D.H. Ingals, "The Smalltalk Graphic Kernel", Byte Vol 6 No 168, August 1981, pp 168–194

J.D. Nicoud, "Video RAMs: Structure and Applications", IEEE Micro, February 1988, pp.8–27

J. Poulton et al., "Implementing a full–scale pixel–planes system", *Proceedings of 1985 Chapel Hill conference on VLSI*, Computer Science Press, 1985, pp 35–60

Ph. Schweizer, "Infographie I et II", Presses Polytechniques Romandes, 1987, 850p.

Ph. Schweizer, "Architecture graphique 3–D parallèle", Thèse de l'Ecole Polytechnique Fédérale de Lausanne No 746, 1988, 290p.

I.E. Sutherland, "Sketchpad: A man–machine Graphical Communication System", SJCC, Spartan Books, Baltimore, 1963, pp.329

C.P. Thacker et al. "ALTO: A Personal Computer", in *Computer Structures: Principles and Examples*, D. Sieviorek et al. editors, McGraw Hill, 1982, pp.549–572

2

Algorithmic Geometry

Th. M. Liebling and A. Prodon
Ecole Polytechnique Fédérale de Lausanne

2.1. Introduction

Generating and displaying complex objects and scenes, making them undergo continuous or discontinuous transformations to simulate motion, growth, etc. requires efficient algorithms handling suitably structured data. The design and analysis of such data structures and algorithms is the subject of what has become to be known as computational or algorithmic geometry. Although the geometrical sets considered there are usually infinite, they are given by a finite number of parameters which is why the ensuing problems are mainly of discrete nature and thus accessible to the notions developed in discrete mathematics. Yet, in opposition to e.g. combinatorial optimization, where interesting problems are usually np-hard, many important problems in computational geometry can be treated with algorithms of polynomial complexity. In fact, a big challenge of computational geometry lies in turning some of these fast algorithms even faster. Typically, one will try to reduce the computational complexity of an algorithm from $O(n^2)$ to $O(nlog(n))$, where n is the size of the appropriately coded input data. Such an acceleration may well be decisive to enable someone to produce an animated scene on a work station.

Several fundamental problems of computational geometry that keep coming up in applications will be briefly treated here. The interested reader will find a detailed account in Aho *et al* (1985), Edelsbrunner (1987), Meier (1988), or Preparata *et al* (1985). Such problems are

Intersection problems, i.e. determine the intersections of families of simple objects like segments, polygons, rectangles, etc.,

Location problems, i.e. given a subdivision of the plane or space into cells, find the one(s) containing a given point or other object,

Convex hulls, i.e. find the smallest convex set enclosing a given family of points or other objects on the plane or in space,

Proximity problems, i.e. given a cluster of points , find a subdivision of space, such that each cell of the subdivision is made up of those points in space that are closer to one point of the cluster than to the others.

The examples below illustrate some applications.

Simulating normal grain growth in polycrystals

Polycrystals, like for instance ceramics, are composite materials formed of individual monocrystalline grains separated by grain boundaries. They may be thought of as forming complexes of polyhedral cells. Under appropriate conditions, polycrystals display what is called normal grain growth, that is, large grains tend to get even larger, at the expense of the smaller ones, which tend to disappear. In this process energy as measured by total surface area of the grain boundaries decreases. This phenomena can be modelled at various different scales. At the lower end of the scale, one can mention atomistic models of Potts or Ising type, see Righetti (1989), which are akin to neuronal cell complexes. At a larger scale, the microscopic models as reported in Telley *et al* (1987) and Telley (1989) consider a grain as the elementary unit. They make extensive use of generalizations of Voronoi complexes (see below), namely Laguerre partitions and succeed in both approximating observations of plane cuts of actual polycrystals and also in adequately reproducing normal grain growth and the accompanying elementary topological transformations.

Both the statistical analysis of plane cuts of the material and the simulation models make intensive use of the algorithms from computational geometry. In this context it is noteworthy to cite an automated image processing process of the plane cuts, which results in a digitized image. Next a statistical test battery, was developed and applied to both the digitized pictures and to the output from the simulation models, thereby making a validation possible. Figure 2.1 (see colour section) shows a two dimensional Laguerre tessellation representing a plane cut of a polycrystal. The Laguerre tessellation is defined by a family of circles. A cell is given by those points of the plane having smaller power with respect to a given circle than with respect to the others. Simulation proceeds by numerically integrating a differential equation in which the rates of change of a circle's parameters are set proportional to the corresponding energy gradient.

Simulating growth of mycelia

Some mycelia species like for instance mucor spinosus display a two dimensional arborescent growth pattern. Growing out of a single spore, hyphae branch at given points and very quickly tend to occupy a nearly circular region. In an ongoing research project, see Indermitte et al (1990), this process is being studied concurrently through observation in vivo and computer simulation. As was described above in the case of polycrystal growth, both seizing the corresponding observations via image processing and also modelling the process for simulation makes heavy use of algorithms from computer graphics. In fact, after suitable modification, many of the algorithms developed to quantify the polycrystals were directly put to use here. The simulation model was developed starting from a purely stochastic approach, in which hyphae simply branch at random times and are considered as parallel processes growing at a regular random speed stopping whenever they get too close to others. The main ingredient of this simulation algorithm is a very efficient segment intersection test of the kind described below. This very simplistic model was soon made more sophisticated by having hyphae interact with each other through the medium. In fact they secrete an inhibitory substance and are endowed with sensors measuring the concentration gradient of this substance. The substance diffuses into the medium and hyphae grow in the direction of the negative gradient of its concentration. First comparisons between simulations and measurements from actual cultures tend to confirm the validity of the model. Figure 2.2 (see colour section) shows an output from the growth simulation model. Several generations of hyphae are characterized by different colours. Figure 2.3 (see colour section) shows the contour lines of the inhibitory substance concentration and the growing mycelium.

Disjoint paths of minimum total length in the plane.

A general problem encountered in designing VLSI circuits consists in choosing routes for the connections to be established between modules placed on a chip. These connections have to form independent - physically disjoint - electrical nets, whose end points are the given pins associated with the modules. In the simplest form of this routing problem each net contains only two terminals, and we are thus looking for disjoint paths connecting given pairs of terminals.

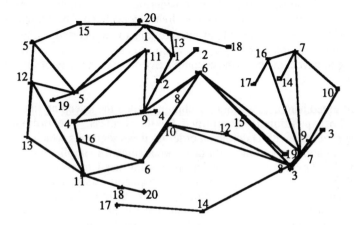

Figure 2.4. Segments used by disjoint paths

Disregarding the other physical constraints of a particular realization, the basic geometric problem is then: given N pairs of points in the plane, determine N paths, each connecting a given pair, such that no two of them cross and the sum of their length is minimum. In this topological formulation each path is an obstacle for the others: two paths may share some points or even some edges, but they will cross if there is no infinitesimal move which makes them disjoint. Such a solution with twenty disjoint paths is shown in Figure 2.4.

Clearly there is an optimal solution in which the paths use only edges of the complete graph constructed on the 2N points, the length of an edge being the euclidean distance between its end points. Hence the problem is combinatorial, but we do not know any polynomial time algorithm to solve it to optimality. Good solutions are obtained efficiently in working on a reduced graph, the Delaunay triangulation on the given points, which already assures that no edges cross.

2.2. Intersections.

Three fundamental problems in computer graphics are to determine if two objects have a non void intersection, construct this intersection, or report all pairwise intersections of N objects. They come up for instance when determining visible parts of a projection of an object described by its faces. Some of the methods commonly used in this domain will be sketched below.

2.2.1. Intersections of segments in the plane.

Problem: given N segments in the plane, each defined by the coordinates of its two end points, report all their intersections.

This problem can of course be solved by considering each segment successively and computing its intersections with each segment not already considered. The complexity of this method is then $O(N^2)$. Since the number K of intersections to be reported can be $N(N-1)/2$, such a method is optimal in the worst case, but when K is small in comparison with N, it would be interesting to have a method whose complexity depends directly of this number K, see Chazelle (1986).

The method presented here uses a general technique called the sweeping line. Consider a vertical line D whose position is determined by its abscissa. The ordinates of the intersections of the segments with D define an order on these segments which remains valid in an infinitesimal neighborhood on the right of D (we assume for simplicity that there is no vertical segment). The key observation which avoids the need to compute the possible intersection of each pair of segments is that, when two segments intersect, they are neighbors in the so defined order for a position of D immediately on the left of this intersection. Thus only the possible intersections of neighbor segments have to be considered in order to find the intersections on the right of D. Now sweeping the plane by moving D from the left to the right will generate all relevant orders of the segments. Fortunately there is only a finite number of positions of D to consider in this process, namely the ones where this order changes, which are the 2N+K abscissas of the segment end points and of the intersections. Moreover, in each such position, only the segments which became neighbors in the new ordering should be tested for a possible intersection. The example in Figure 2.5 shows a position of D, the ordering of the segments valid on the right of the successive positions and the two positions in which intersections are detected.

Two data structures are needed to implement this sweeping line method. The first maintains the positions of D to be considered. Since we move D from left to right to the next interesting position, a priority queue is adequate, which initially contains the abscissas of the point extremities, and in which the intersection abscissas are introduced as they are found. The second one maintains the segments intersecting D ordered along the ordinates of these intersections. According to the change occurring in the current position of D it must support insertion (left extremity of a segment), deletion (right extremity of a segment) and permutation of two elements (intersection point). An appropriate structure is thus a height-balanced tree (AVL-tree), in which these operations are executed in $O(\log N)$ time. The final complexity of this method is then $O((N+K)\log N)$.

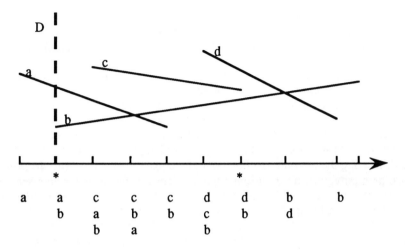

Figure 2.5. Intersection of segments using a sweeping line

2.2.2. Intersection of two convex polygons.

Problem: Given two convex polygons P and Q, each one defined by the list of its vertices in the order they are encountered in a walk around the polygon in the trigonometric sense, determine their intersection.

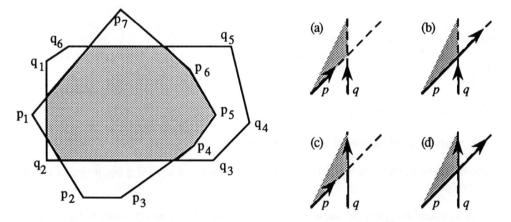

Figure 2.6. Intersection of two convex polygons.   **Figure 2.7.** Advance mechanism.

Here we note in the polygon the edges defining it as well as the interior region they determine. In general, the intersection of two polygons is not connected: two star-shaped polygons, each having $O(N)$ vertices, may already have an intersection composed of $O(N^2)$ connected components. But the intersection of two convex polygons, P with vertices $p_1,...,$ p_r and Q with vertices $q_1,...,q_s$, is itself a convex polygon, having at most $r+s$ vertices. Each such vertex being the intersection of an edge of P with an edge of Q, our aim is to determine all these intersections, scanning successively each edge in a walk around each polygon. Here too the convexity property can be exploited in a walk mechanism which avoids having to consider all possible pairwise intersections and leads to an algorithm in O(r+s) time. The principle of this algorithm will be explained on the example in Figure 2.6.

The boundary of P ∩ Q is composed alternatively of chains of P and chains of Q, a chain being possibly just a portion of an edge. Walking around the polygons counterclockwise, two consecutive vertices of P ∩ Q are joined by two chains, one belonging to one of the given polygon and to the boundary of the intersection, the other one to the ·other given polygon and located entirely to the right of the first one. These two chains define a sickle. Having located a first intersection point, say p_3p_4 with q_2q_3 , we will advance on these two chains until locating the next intersection point, here p_6p_7 with q_5q_6 , with the principle not to advance on a polygon whose current edge may contain the searched intersection point. There are essentially four possible cases for the intersection of the two current edges p and q , which are illustrated in Figure 2.7: in case (c) and (d) q may contain the intersection point so we advance on P, while in case (a) and (b) p may contain it and we advance on Q. So we would in the example successively advance from p_3p_4 and q_2q_3 (case d) to q_3q_4 (crossing the current edge of the other polygon as in case a), then q_4q_5 (c), p_4p_5 (c), p_5p_6 (a), q_5q_6 (c), p_6p_7 (d), so reaching the end point of this sickle. After most r + s advance steps along these rules we will have scanned all the edges and hence determined all the intersection points.

The first intersection point is found - if it exists - using the same rules: If the two starting edges belong to the same sickle then we will advance possibly on both polygons until finding the second (and last) intersection point on this sickle, or else we will advance only on one of the polygons, depending on the relative position of the starting edges, until reaching an edge in the sickle to which the unchanged edge belongs. Thus after most r + s advance steps we will either have found a first intersection point or shown that none exists, since all the edges of one polygon will have been scanned by that time. In the last case either P is contained in Q, or Q is contained in P, which can easily be tested as shown in the next section, or their intersection is void.

2.3. Location problems.

2.3.1. Location of a point in a polygon.

Problem: given a simple polygon P in the plane, determine if a point q lies inside or outside this polygon.

For a general polygon, the answer is deduced from a well known theorem of Jordan which says that a closed curve partitions the plane in two parts, its interior and exterior. Thus we count the number of intersections of a half-line (without loss of generality a horizontal line), issued from q, with the contour of P: the point q lies outside P if this number is even, inside P if it is odd. Counting the number of intersections is done by intersecting each edge defining P with the half-line, which requires O(N) calculations when P has N edges.

One way to handle the degenerate cases consists in considering only the non horizontal edges of P and defining for these the lower extremity to be excluded from the edge and the upper extremity to be part of it. Thus in the example in Figure 2.8 this number of intersections is 5, an odd number, so q lies inside P. The case where q lies on the boundary of P is trivial.

When the polygon is known to be convex and is given by the ordered list of its vertices, the complexity of the inclusion test can be reduced to O(logN). For this, the plane is partitioned into sectors defined by two consecutive vertices of P and any interior point O (any convex combination of three vertices of P produces such a point). A binary search locates point q in one of these sectors, say p_iOp_{i+1}, then a single test determines on which side of p_ip_{i+1} point q lies (see Figure 2.9). No angle manipulation is necessary in these operations since the relevant information is the sign of an oriented triangle, namely triangle Op_iq in the binary search and triangle $p_ip_{i+1}q$ in the last test, which is the sign of the 2x2 determinant of the matrix with columns $p_{i+1} - p_i$, $q - p_i$.

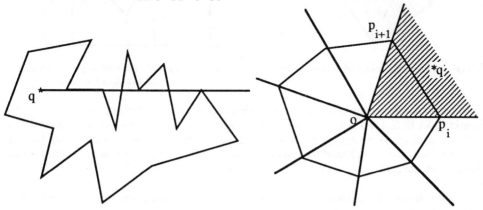

Figure 2.8. Location of a point in a simple polygon.

Figure 2.9. Location of a point in a convex polygon.

This method can also be applied for a star polygon, a polygon which is not necessarily convex but has an interior point from which any point of the polygon is visible.

2.3.2 Location of a point in a monotone subdivision of the plane.

Problem: Given a partition of the plane into cells or polygonal regions, determine the region where a given point q lies.

Clearly one way to solve this problem is to test for the inclusion of q into each polygon defined by the partition, but if the problem has to be solved for many points with the same partition, some preconditioning can reduce the search time considerably.

A region is said to be monotone with respect to a given direction, which without loss of generality will be assumed horizontal, if its intersection with any line perpendicular to this direction is a single interval.

A subdivision of the plane is monotone if all its regions are monotone. This is equivalent to require that in any vertex of the subdivision there is at least one edge incident to it from the left and at least one from the right. If this property is not initially fulfilled, it can be achieved by refining the subdivision, that is subdividing a region containing a vertex with missing edges (see Figure 2.10). This can be done in O(NlogN) time, the complexity for sorting the abscissa of the N subdivision's vertices, and without notably increasing the number of edges which remains linear in the number of vertices as in any planar graph. Thus from now on the given subdivision is assumed monotone.

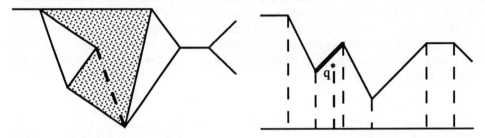

Figure 2.10. Subdivision of a non monotone region.

Figure 2.11. Location of a point with respect to a monotone chain.

In particular the edges of a monotone chain are naturally ordered, say from left to right. Thus in order to test if a point lies above or below a monotone chain one can simply locate the abscissa of the point among the sorted abscissas of the extremities of the edges forming this chain by binary search, and then decide if the point is above or below the corresponding edge (see Figure 2.11).

In order to use this property a complete system of N-1 monotone chains describing the given subdivision is constructed, such that:

- the chains use only edges of the monotone subdivision,
- the chains do not cross and hence are ordered with respect to the relation "above",
- there is exactly one region of the subdivision between two consecutive chains in that ordering, namely region R_i between the two chains s_i and s_{i+1}.

These N-1 monotone chains thus define the same regions as the given subdivision (see Figure 2.12), but locating a point in a region now reduces to locating the point between two consecutive chains, which is achieved by a binary search.

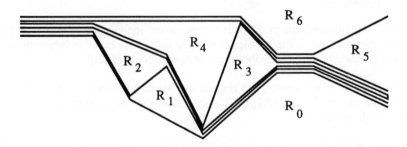

Figure 2.12. A complete system of monotone chains

Although quite interesting, this idea of formulating an equivalent subdivision with monotone chains is not   sufficient to produce an efficient algorithm. In the above formulation we would need to store N-1 chains, each composed of O(N) edges, thus a memory requirement of $O(N^2)$ units; moreover the location of a point would require comparing it to log(N) chains, each comparison involving O(logN) comparisons of abscissas. Clearly there are too many redundancies in this formulation, which may be eliminated using adequate data structures. In the binary decision tree whose nodes represent the (comparison with) monotone chains, each edge of the subdivision should be stored only once instead of N-1 times, since when the point has been located with respect to such an edge, the result of that comparison is known and has not to be computed again deeper in this tree. Also, once the point abscissa has been located in an X-interval occurring in a chain associated with a node in the decision tree, this information should be used for the next comparison in the corresponding son of this node.This is achieved by refining these X-intervals such that each X-interval occurring in a node intersects at most two of the X-intervals occurring in each of its two sons. Then at most one test will determine which edge the point has to be compared with in order to be located with respect to the next chain. This refinement of the X-intervals associated to each node of the decision tree, starting from the leaves and moving to the root, keeps the total number of these intervals linear in N (it can be shown to be at most 4N). The location of the point's abscissa thereby reduces to a single binary search with respect to the intervals associated to the root node.

A rigorous implementation of these ideas leads then to an optimal algorithm described in Edelsbrunner *et al* (1986), using O(NlogN) time units for preconditioning, O(N) memory units and O(logN) time units for the location of any point.

2.4. Convex hulls.

A set is convex if the segment joining any pair of its points is entirely contained in it.The convex hull of a set of points is the smallest convex set containing them all.

There are many known methods to determine the convex hull of a set of points in the plane; we shall sketch one of them, of the iterative type, which can be generalized to higher dimensions. It can also be modified to maintain the convex hull of a set in which points are dynamically inserted or deleted. In this simple version all the points $p_1, p_2,..., p_N$ are known from the beginning and numbered according to growing abscissas. An iteration goes as follows: suppose we have already constructed the convex hull C_{i-1} of the points $p_1,...,p_{i-1}$ and that C_{i-1} is stored as the sorted list of its vertices,when turning counterclockwise. In order to add p_i and construct C_i we determine the two lines p_iv_b and p_iv_t issued from p_i and supporting C_{i-1}. Thereby C_{i-1} decomposes then into two chains with extremities v_t and v_b. We eliminate the one of these chains which will no longer belong to C_i (the one containing p_{i-1}) and replace it by v_tp_i and p_iv_b, that is we introduce p_i in the list of vertices and let p_i be the successor of v_b and v_t the successor of p_i.

Two supporting lines are constructed by turning around C_{i-1} and considering its vertices , starting from p_{i-1} which is its right most point . Point v_t is the first point encountered when turning counterclockwise such that p_i lies strictly below the line passing through v_t and the successor of v_t. Analogously v_b is the first point encountered when turning clockwise such that p_i lies above the line through v_b and the predecessor of v_b. This construction is illustrated in Figure 2.13. Since each point considered in such an iteration either produces a supporting line or is eliminated, the time needed for the N iterations remains linear in N, and the algorithm's complexity is O(NlogN) accounting for the preliminary sorting of the abscissas.

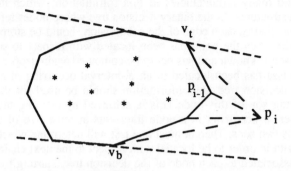

Figure 2.13. Construction of the convex hull.

2.5. Proximity problems.

Let $S := \{p_1, p_2,..., p_N\}$ be a set of N given points in the plane The Voronoi cell of a point p_i is defined as the set of all points of the plane which are nearer to p_i than to any other point of S. Such a cell is the intersection of the half-planes containing p_i and delimited by the bisectors between p_i and the other points of S, and thus a convex polygon. The set of the N Voronoi cells associated with the N given points partitions the plane and is called a Voronoi diagram. The dual of this diagram, obtained by joining all pairs of points whose Voronoi cells are neighbors, is a triangulation called the Delaunay triangulation of the set S. It partitions the convex hull of S in triangles having the points as vertices and containing none of the points in their interior. Figure 2.14 shows such a triangulation and the corresponding Voronoi diagram.

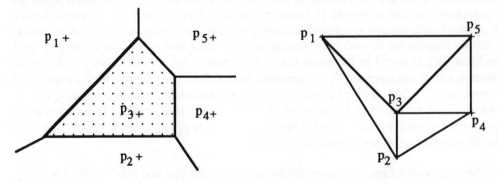

Figure 2.14. Voronoi diagram and Delaunay triangulation of a set of points

Voronoi diagrams and Delaunay triangulations are fundamental instruments used in the solution of many combinatorial problems related with distance evaluations, such as:

- Two nearest neighbors: Determine two points in S whose distance is minimum.
- All nearest neighbors: Determine a nearest neighbor of each point of S .
- Euclidean minimum spanning tree: Determine a spanning tree on the points in S having minimum total length, the length of an edge being the euclidean distance between its extremities.
- Nearest neighbor of an arbitrary point: Given any point q in the plane, determine the point of S nearest to q.
- k-nearest neighbors of an arbitrary point: Given any point in the plane, determine k points of S which are nearer to q than any other point in S.
- Minimum weighted matching: Assuming that N is even, partition S into N/2 pairs such that the total length of the resulting line segments is minimum.
- Minimum euclidean travelling salesman tour: Determine a tour passing through each point of S and having minimum length.

For the first three problems it can be shown that it suffices to consider the edges of the Delaunay triangulation to solve them, thus considering at most 3N-6 edges instead of N^2 .

The next two problems are search problems for which the search time can be reduced when preprocessing is allowed, the preprocessing consisting in this case in the construction of a Voronoi diagram, the search reducing then in locating the point in a cell of the partition of the plane thus obtained.

The last two problems are examples of classical combinatorial optimization problems with data restricted here to euclidean distances. In that case, in opposition to the minimum spanning tree which is also such an optimization problem, looking for a solution on the Delaunay triangulation does not always produce an optimal solution. But the experience shows that admissible solutions found in that way are almost always very good, often optimal, and always found efficiently.

One way to construct a Delaunay triangulation is to use a divide and conquer method. This is one of the most widely used general method in computational geometry. Its principle is to fragment a problem into subproblems of the same kind, solve the subproblems recursively and combine their solutions into a solution of the original problem. In our case this means separate S into two sets S_1 and S_2 of equal size, recursively construct the Delaunay triangulations D_1 of S_1 and D_2 of S_2 and then in a merging step construct D of S_1 $\cup$ S_2 . A linear separation is obtained in constant time when the points are previously sorted along growing abscissas, and if the last step can be executed in linear time with respect to the number of points in $S_1 \cup S_2$, the total construction time remains in O(NlogN), which can be shown optimal.

This merging step is illustrated in Figure 2.15. The triangulation D consists of some edges of D_1 and D_2 and new edges having one extremity in S_1 , the other in S_2 , constructed iteratively from bottom to top as a kind of ladder, starting with the lower supporting line, p2p5 in the example, which is certainly part of the convex hull and thus of the Delaunay triangulation. The main instrument to decide which edges belong to D is a test deciding if a circle through three points contains a fourth point or not, resulting from the property of a

Delaunay triangle to contain no other point of S than its vertices. In the example, having found p_2p_6 as an edge of D, the next triangle will have this edge as its base. We grow a circle passing through p_2 and p_6 until it meets another point of S. This point will be the extremity of an edge of D_1 (or D_2) issued from p_2 (respectively p_6). In our case it is p_3 , thus the edge p_2p_4 does not belong to D and is eliminated. This process is repeated, from p_3p_6 and so on, until reaching the above supporting line p_4p_7 .

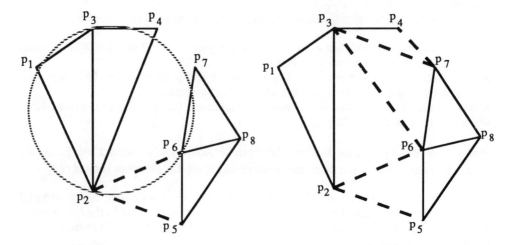

Figure 2.15. Merging step in the construction of the Delaunay triangulation

The growing circle process is implemented by successively considering the edges issued from p_2 in D_1 in the counterclockwise sense, until a potential candidate is found. It is an edge e such that the extremity of its successor is not contained in the circle through the vertices of the triangle it forms together with p_2p_6. A second potential candidate is searched analogously in D_2 turning around p_6 in clockwise direction. Then the circle test is applied to both candidates, here p_3p_6 and p_2p_7 , determining the next edge p_3p_6. This can be done efficiently using appropriate data structures for the representation of the triangulations which are planar graphs. The Quad-edge described in Guibas et al (1985) is such a structure that allows us to find the successor and the predecessor of each edge directly when turning around one of its extremities or around one of the faces on each side. Thus the Quad-edge constitutes simultaneously an encoding of a planar graph and its dual, in our case given by the Delaunay triangulation and the Voronoi diagram.

References

Aho, A.V., Hopcroft, J.E., and Ullman, J.D. (1985). Data Structures and Algorithms, Addison-Wesley.

Chazelle, B. (1986). Reporting and counting segment intersections, J. Comput. System Sci. 32, 156-182.

Edelsbrunner, H. (1987). Algorithms in Combinatorial Geometry. Springer-Verlag, Berlin.

Edelsbrunner, H., Guibas, L.J., and Stolfi, J. (1986). Optimal point location in a monotone subdivision, SIAM J. Comput. 15, 317-340.

Guibas, L.J., and Stolfi, J. (1985). Primitives for the manipulation of general subdivisions and the computation of Voronoi diagrams, ACM Trans. Graphics 4, 74-123.

Indermitte, C., Liebling, Th.M., and Clemençon, H. (1990). Simulation de la croissance du mycelium, Rapport 900115, EPFL et UNIL, Lausanne.

Meier, A. (1988). Geométrie algorithmique, in Infographie et Applications (Eds. T. Liebling and H. Röthlisberger) pp. 159-194, Masson, Paris.

Preparata, F.P., and Shamos, M.I. (1985). Computational Geometry - an Introduction. Springer-Verlag, New York.

Righetti, F., Liebling, Th.M., and Mocellin, A. (1989). The use of large scale Potts pseudo-atomic model for three-dimensional grain growth simulation, Acta Stereologica 8/2.

Telley, H. (1989). Modélisation et simulation bidimensionnelle de la croissance des poly-cristaux. Thèse 780, EPFL, Lausanne.

Telley, H., Liebling, Th.M., and Mocellin, A. (1987). Reconstruction of polycrystalline structures: A new application of combinatorial optimization, J. Computing 38, 1-11.

3

Surface Visualization*

Peter Buser, Boris Radic and Klaus-Dieter Semmler
Ecole Polytechnique Fédérale de Lausanne

3.1. The Quest for Realism

When does a surface look "good"? In computer graphics there are a number of techniques which improve the quality of an image, using *texture, specular light*, and *ray tracing*, etc. These techniques improve the realism of an image. For instance, if an object like a car body is to be shown on the screen, then it should look the way we know it looks. Science and engineering, on the other hand, produce objects which do not belong to every day life. These objects do not have to look real. (What does "real" mean for the diagram of a temperature distribution?) In science, the purpose of a picture is to visualize complicated phenomena. But if a picture wants to help us understand a phenomenon, we must understand the picture. Scientific drawing disposes of numerous *picture symbols* which we have learned to read. A few such symbols are shown in Fig. 3.1 on the next page.

The first example might be a part of an electric circuit. Then comes a cube, followed by a sphere, an annulus and a torus. Why does the second example look three dimensional but the first does not? Why does Figure d) seem to be in the plane, whereas Figures c) and e) are in space? The answer is that we have *learned* to see it this way: the dots are the symbol for being hidden; interrupted edges contain depth information, etc. We may thus say that a scientific picture looks "real" when the relevant picture symbols are present.

While, for hand drawings such as in Fig. 3.1, the useful symbols were found long ago, computer graphics is in a process of creating and testing appropriate representation techniques, and it is not known yet which will be the best.

Here are some criteria for the visualization of surfaces.

* The shape should be easy to recognize.
* Enhancing attributes must not dominate the nature of the surface itself.
* Additional properties (such as curvature) should be visualizable.

---

* This work has been supported by the Swiss National Science Foundation

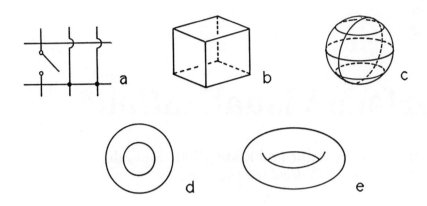

Figure 3.1. Picture symbols

In the following pages we present some concepts and ideas which we find useful. Part of the material is standard, and part of it belongs to current research. The small fragments of programs included in the text are written in Pseudo-C code.

3.2. What Is a Surface and How Is It Stored?

3.2.1. Level Surfaces

A surface which arises frequently in science and engineering is the so-called *level surface*. A typical case is the temperature in a room, usually represented as a function $t = f(\mathbf{x})$ where t is the temperature and $\mathbf{x}$ is a variable point in space with coordinates x_1, x_2, x_3. The points at which the temperature assumes a given value, say c, trace out a surface in general, given by the equation

$$t = f(\mathbf{x}),$$

the *level surface* of f with respect to c. The temperature in the room can then be *visualized* by showing the level surfaces for different values of c. A classroom example for this is the function $f(\mathbf{x}) = x_1^2 + x_2^2 + x_3^2$ whose level surfaces are given by the equation

$$x_1^2 + x_2^2 + x_3^2 - c = 0.$$

For $c > 0$, these level surfaces are spheres of radius $\sqrt{c}$.

3.2.2. Unstructured Discrete Surfaces

To draw a level surface, one needs numerical data. One approach is to compute sufficiently many points on the surface and to consider these points as "the surface". When a surface comes in this form, we call it *unstructured discrete* : *discrete* because it consists of finitely many points, *unstructured* because no further information about the relative position of the points is given. It is obtained in the following steps.

- Divide the part of space, which is to be represented, into small cubes parallel to the coordinate axes.
- Test for each edge of each cube whether it intersects the surface.
- If it does, compute and store (or draw) the intersection point.

The test and the computation of the intersection point for the level surface "$f(x) - c = 0$" are as follows. Consider the end-points **a**, **b** of an edge of a cube. If $f(a) \leq c$ and $f(b) > c$, then there must be a point **x** somewhere between **a** and **b** where $f(x) = c$. The same conclusion holds if $f(a) \geq c$ and $f(b) < c$. Inversely, if the cubes are small and if f is not too exceptional, then the edge does not intersect the level surface if both $f(a)$ and $f(b)$ are above or below c. One computes therefore the product

$$\sigma_{ab} = (f(b) - c)(c - f(a)).$$

If $\sigma_{ab} \geq 0$, then the test is positive, and the procedure continues by computing the point where the edge intersects the surface. For this, linear interpolation is sufficiently accurate, in general. The coordinates of the intersection point x are then given by the formula

$$x_k = \frac{a_k(f(b) - c) + b_k(c - f(a))}{f(b) - f(a)}, \quad (k = 1, 2, 3),$$

where a_k and b_k are the coordinates of **a** and **b**. (If $f(a) = f(b) = c$, take **x** to be the midpoint between **a** and **b**.)

Fig. 3.9 and 3.10 (see colour section) show an example obtained in this way and intensified with SFUMATO.

3.2.3. Parameterized Surfaces

Dynamical processes lead to another type of surface: the *parametric* or *parameterized surfaces*. Formally, they are given by a triplet of equations

$$\begin{aligned} x_1 &= r_1(u, v) \\ x_2 &= r_2(u, v), \quad (u, v) \in D, \\ x_3 &= r_3(u, v) \end{aligned}$$

where u, v are the *parameters*, x_1, x_2, x_3 are the coordinates of a point x in space, and r_1, r_2, r_3 are functions. The triplet of equations is sometimes written in an abbreviated form as

$$x = r(u, v), \quad (u, v) \in D.$$

D is the parameter *domain* of the surface and is a subset of the plane. Fig. 3.2 shows an example: a catenoid is isometrically deformed into a helicoid. The equations are

$$\begin{aligned} x_1 &= \cos \alpha \cosh u \cos v + \sin \alpha \sinh u \cos v \\ x_2 &= -\cos \alpha \cosh u \sin v + \sin \alpha \sinh u \sin v \\ x_3 &= u \cos \alpha + v \sin \alpha \end{aligned}$$

The constant α can be varied. For $\alpha = 0$ we obtain the initial surface, the catenoid; for $\alpha = 2\pi$ we get the final surface, the helicoid.

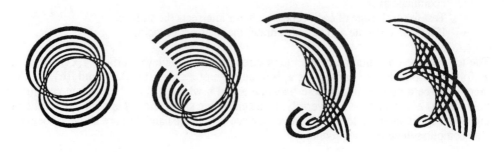

Figure 3.2. A catenoid is isometrically deformed into a helicoid

3.2.4. Structured Discrete Surfaces

An unstructured discrete surface is a finite number of points, usually stored in a data list. Together with each point, one may store further information about the surface, for instance the normal vector (if known), the color, etc. The stored surface then contains more information, but it is still unstructured and therefore only accessible to those drawing routines which deal with points. If we want to apply routines which deal with segments or with polygons, the data list must be *organized* in such a way that we can tell which families of points together form a segment or a polygon (or perhaps a more complicated aggregate). If a surface comes with an organized data list, we call it *structured discrete*.

A simple example is the array of $N*M*3$ real numbers $r[N][M][3]$ which is interpreted as a set of $N*M$ points in space "organized" in the following way: for each i and j the four points $r[i][j]$, $r[i+1][j]$, $r[i+1][j+1]$ and $r[i][j+1]$ together form a quadrangle.

In general, a structured discrete surface is stored as a data list (as in the unstructured case) together with a so-called *adjacency matrix* which describes how the points in the list are grouped into segments or polygons.

3.2.4.1. To Structure Parameterized Surfaces

Assume that a surface is given by the parametric equation $x = r(u, v)$ with u ranging from u_{min} to u_{max} and v ranging from v_{min} to v_{max}. In this case the domain D of the surface has a rectangular shape. A structured discrete form is obtained by splitting up the range of u and v into M intervals of size $\Delta u = (u_{max} - u_{min})/M$ and N intervals of size $\Delta v = (v_{max} - v_{min})/N$. One then stores the points with coordinates $u_{min} + i*\Delta u$ and $v_{min} + j*\Delta v$, ($i = 0, \ldots, M - 1; j = 0, \ldots, N - 1$).

In Section 3.2 we describe how this procedure can be extended, without further modification, to non-rectangular domains, simply by setting a certain *indicator* to be 0 or 1.

3.2.4.2. To Structure Level Surfaces

In Section 3.2.2 we showed how to compute points on a level surface S given by the equation $f(\mathbf{x}) - c = 0$. We now describe a way to organize these points into polygons (triangles or quadrangles).

Let us consider one of the cubes Q in the space decomposition of Section 3.2.2 such that Q intersects S. If Q is sufficiently small, then the intersection points of its edges with S form a polygon. In order to bypass the problem of how to enumerate the vertices of this polygon in consecutive order (cf Magnenat-Thalmann, 1987), we further decompose Q into tetrahedrons as shown in Fig. 3.3.

Fig. 3.4 illustrates how a given tetrahedron T may intersect S. The $\oplus$'s mark the vertices where $f > c$, and the $\ominus$ marks the vertices where $f \leq c$. There are only three possibilities:

a) two vertices of one sign and two vertices of the other,
b) three vertices of one sign and one vertex of the other,
c) all vertices of the same sign.

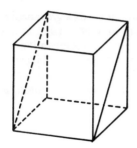 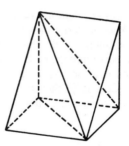

Figure 3.3. A cube is decomposed into 6 tetrahedrons.

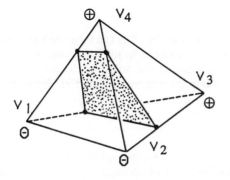 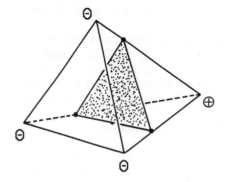

Figure 3.4. Intersections with a level surface

In case c) the test of Section 3.2.2 is negative for all edges, and T does not intersect S. (There are some exceptional cases in which this is not correct; these cases can usually be ruled out by taking a decomposition with smaller cubes.)

A moment's reflection shows that there is essentially only one way to distribute two $\oplus$'s and two ¨'s on the vertices of T. Also, there is only one way to distribute three $\oplus$'s and one ¨, or three ¨'s and one $\oplus$. Hence, Fig. 3.4 exhausts all possible situations.

In case a) the intersection points determine a quadrangle, and we must order them in such a way that no pair of consecutive points lies on a diagonal. This can be achieved as follows. Label as v_1 and v_2 the vertices of T with the ¨ and as v_3 and v_4 those with the $\oplus$. Then take the intersecting edges in this order: $v_1 v_3$, $v_2 v_3$, $v_2 v_4$, $v_1 v_4$ and enumerate the points of intersection accordingly.

The triangles and quadrangles obtained from the tetrahedrons of the decomposition can either be stored in a list or memorized by an appropriate adjacency matrix.

3.3 How to Draw a Surface and How to Recognize Its Shape

Of course this problem has no single solution. The answer depends a great deal on whether an object is known to the observer or not. Thus, a picture may be "clear" to one person and puzzling to another. We should also distinguish between the visualization of *real* objects which are best represented in their natural environment, and *abstract* objects which must be represented in a form such that scientific data are visible. In this chapter we concentrate on the latter. Certainly, the best thing to do is to use more than one method, and to switch interactively from one to another.

3.3.1. The Tool Kit

Here is a list of tools a graphics package should contain. Some of these tools are standard, some are in a state of development, and some are simply ideas.

3.3.1.1. Animation and Viewing

These are the basic techniques: zoom, rotate and translate the object or the observer in space. Use light sources. Add appropriate picture symbols which help to distinguish between the *apparent* deformation caused by a change of the position and an actual deformation of the surface (such as the deformation of the catenoid into the helicoid in Fig. 3.2). Show the axis of rotation. Indicate the present position.

3.3.1.2. Transparency

When several surfaces are shown at the same time, or if a surface is complicated, then large parts may be hidden by others. In this case one likes to look through the surface by making it transparent, for instance by appropriately adding the colors of overlapping patches. Most of today's hardware does not yet support this. In Section 3.4 we present a method which yields transparency in a different way based on the concept of *density*.

3.3.1.3. Stripes

Another method to look through a surface and detect regions of interest is to cut away parts, for instance stripes, which are adapted to the parameter lines as shown in Fig. 3.2. To obtain a surface in this form, it is not necessary to really cut, it suffices to tell the drawing routine *not to show those parts*. For this, we need an *indicator* for each surface element which may be set to 1 ("show") or 0 ("don't show").

3.3.1.4. Cutting

Set a volume in space (such as a small ball, a cylinder, a half space, etc.) and cut away all parts of the surface which are outside this volume. This form of cutting is quite different from the previous one and can be used to scan through space in order to find particular parts of a surface.

3.3.1.5. Thinning

Reduce the number of parameter lines without reducing the resolution: a structured discrete surface resulting from the procedure of Section 3.2.4.1 and shown for instance in wireframe, does not look smooth unless the intervals Δu and Δv are small (meaning high resolution). Pictures then may show too many parameter lines. In order to reduce the number of parameter lines, one does not need to give up the resolution, it suffices *not to show* all the lines. This can be realized with an indicator as for the drawing of stripes.

3.3.1.6. Thickening

Add some thickness to a surface: Borderlines are difficult to recognize, even when the color of the surface is different from that of the background. This can be improved by thickening the surface so that the border line becomes a *band*. Fig. 3.5 indicates a way to realize this for parameterized surfaces. The polygons at the boundary of the surface are replaced by thin wedges. A thickening can also be obtained with SFUMATO (cf. Section 3.4.3.4).

3.3.1.7. Contours

Apparent contours do not only appear at the border of the image of the surface, they appear also at all those points where the tangent plane is parallel to the viewing direction. On real objects these are the points where the human eye has difficulties in focusing, an enhancement of the contours improves therefore the realism of a picture. It can be obtained via the concepts of density, as in SFUMATO, or by differential geometric methods.

3.3.1.8. Two-Sidedness

The visual perception of a surface can be simplified when different colors or different textures are used on the two sides of the surface. (This would be somewhat difficult on the Möbius band, though). Two-sidedness can be implemented by working with the pair {N, -N} where N is the normal vector.

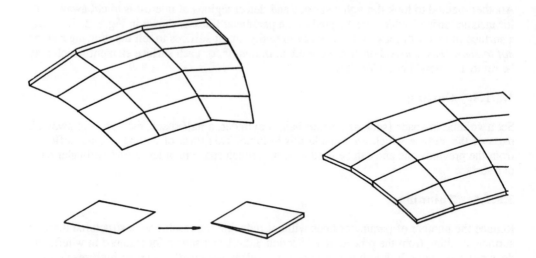

Figure 3.5. Wedged polygons at the boundary yield depth information

3.3.1.9. Curvature.

An important attribute of a surface is its curvature. A real object which is folded (for instance a credit card...) shows signs of stretch as a function of the curvature. An appropriate texture or a change of color can therefore be used as a picture symbol to indicate the curvature of a surface.

3.3.2. Ways of Presentation

The implementation of the previous tools depends on the structure of the surface, that is, on how the data are organized. There are essentially four types of data in scientific representations.

- Point-like data as for unstructured discrete surfaces.
- Line-like data as for space curves or grids of parameter lines.
- Polygon-like data as for the triangles and quadrilaterals of level surfaces.
- Polyhedron-like data as in crystal structures.

A surface which is presented in one of the first three types is said to be in point, line, or polygon mode respectively. The fourth type does not apply to surfaces and has been mentioned for completeness.

The philosophy is the following. A surface in any of the three modes is made up of *surface elements*. These elements are respectively *points*, straight line *segments*, or *polygons*. The vertices of a polygon need not be contained in the same plane. Intersections, visibility and transparency are handled by the z-buffer.

Along with each surface element come a number of additional data, stored in a *paint box*. Each surface element has its own paint box.

3.3.2.1. Point Mode

Here we put the following attributes into the paint box

- The indicator to switch the point on or off.
- The normal vector, if present.
- Scalar functions such as curvature, temperature, etc.

A routine for this which calls for the drawing hardware may look as follows.

```
draw_points (r, paint_box)
3D data r;
paintbox_structure  paint_box;
{
  if( paint_box.indicator == 0)   return;
  if( use_of_normals == shading)   color(paint_box.normal);
  if( use_of_scalars == curvature)   color(paint_box.scalar1);
  pixel(r);
}.
```

3.3.2.2. Line Mode

The oldest drawing routine for surfaces in computer graphics is *wireframe*. It gives a fair picture of the surface but its main setbacks are the lack of depth information. A more sophisticated form of wireframe can be obtained by using surfaces which are in line mode. Here are some options which a user may want to find in the paint box.

- Show/do-not-show the segment.
- Use dotted lines for hidden segments.
- Do not show hidden segments.
- Add halo.

To *add halo* means to slightly blur or interrupt an edge when it passes behind another edge. Such interruptions are shown in Fig. 3.1 and 3.3. We do not know as yet how to realize this effect by means of a z-buffer. (For an algorithm which interrupts segments see (Lucas, 1988) The same holds for the drawing of dotted lines.

3.3.2.3. Polygon Mode

This is the basic mode for the drawing of surfaces. The assumption is, of course, that the surface is structured discrete. We begin again by listing some of the options which one may want to find in the paint box.

- Show/do-not-show the polygon.
- One-sided/two-sided.
- Show boundary of polygon only.
- Show boundary and interior in different colors.

- Change color if normal vector is almost orthogonal to observer.
- Enhance contour lines.
- Decrease density.
- Indicate the curvature.

Exercise. Which of the above options can you find in the following routine?

```
draw_polygon(N, points, paint_box)

int N   /* number of points, usually 3 or 4 */
3D data points[N], normals[N];
paintbox_structure   paint_box;
{
    float       temporary, temporary.old, temp_normal[3];
    if( paint_box.indicator == 0) return;
    beginpolygon();
    for( i=0; i < N ; i = i + 1 )
    {
        temp_normal = paint_box.normal[i];
        temporary = < temp.normal, view_vector >;
        if( use_of_normals == shading)
        {    if ( temporary > 0 )
               {
                    change_sign(temp_normal);
                    if (two_sided)     backsidematerial();
                    else               frontsidematerial();
               }
            if (contouring)
               {
                    if (temporary*temporary.old <= 0 )
                            color( contour);
                    else    color(temp_normal);
                    temporary.old = temporary;
               }
        }
        else if( use_of_scalars == curvature)
                color(paint_box.scalar1);
        vertex(points[i]);
    }
    endpolygon();}
Fill_indicators()
```

The options in the above list should speak for themselves, and we move on to the applications.

Stripes. This is the algorithm which cuts out the stripes of Section 3.2.1.3. For simplicity we assume that the polygons (quadrangles) which belong to the $N*M$ grid in the domain D are enumerated as $P[i][j]$ with i ranging from 0 to $M-1$ and j ranging from 0 to $N-1$. If i is odd we set indicator($P[i][j]$) = 1, and if i is even we set indicator($P[i][j]$) = 0.

Light Sources and Normal Vectors. A surface drawn with filled polygons with a single constant color would yield a silhouette. A better view is obtained by using light effects. For this a *normal vector*, say N, of the polygon has to be known. We refer to

Section 3.5.1.1 for the computation of **N**. We emphasize that each polygon has *two* normal vectors ±**N** pointing in opposite directions. For shading algorithms to work properly, one needs the one pointing to the observer, so we should always begin with the following instruction.

If **N** *points away from the observer, change the sign of* **N**.

We now consider a given polygon (i.e. a surface element). To avoid any misinterpretation, we denote here by **N'** the normal vector of the polygon which points to the observer. If the polygon is opaque (which is usually the case), then the observer cannot see any brightness issuing from a light source which is located *behind* the patch. Using Lambert's law we define therefore the *observed brightness* B_{obs} of the polygon with respect to a given light source via the formula

$$B_{obs} = \begin{cases} B_0 \cos \varphi & \text{if} \quad 0 \le \varphi_{N'L} \le \pi/2 \\ 0 & \text{if} \quad \pi/2 \le \varphi_{N'L} \le \pi \end{cases}$$

where $\varphi_{N'L}$ is the angle between **N'** and the vector which points from the polygon to the light source. B_0 denotes the intrinsic "color"-value of the polygon.

Two-Sidedness. If the polygon has two colors, say B_1 on the side of **N** and B_2 on the side of -**N**, then we substitute (in the definition of B_{obs}) B_1 for B_0 if **N'** = **N**, and we substitute B_2 for B_0 if **N'** = -**N**. The reader may check that with these conventions the two-sided polygon is dark if the light source is *behind* it.

3.4. SFUMATO

3.4.1. What is SFUMATO?

The name "sfumato" (from Italian, *sfumare*, to shade off) is borrowed from a painting technique used by the Italians of the Renaissance. It is defined it as "the gradual transition between areas of different color, avoiding sharp outlines". In computer graphics we introduce it as a drawing routine, based on clouds, which gives a certain fuzzy appearance of the objects. One of the advantages of this is that we can physiologically interpolate unstructured discrete data. The following figure shows the idea in the case of a curve. The example is a set of points which may come from a curve. The data are unstructured, and seemingly difficult to order consecutively. Yet, the human eye has the tendency to integrate the data and to see a curve. In order to enhance this effect, we replaced each point by a cloud. SFUMATO is technique which applies this to surfaces.

3.4.2. Clouds

We now discuss how clouds are built. Here is an example:

$$\text{Cloud}(x_0, \text{density}, v_1, v_2, v_2) = \{ x_i \mid x_i = x_0 + (\text{rand}() - \frac{1}{2})*v_1 +$$
$$(\text{rand}() - \frac{1}{2})*v_2 + (\text{rand}() - \frac{1}{2})*v_3 , i = 1..\text{load} \}.$$

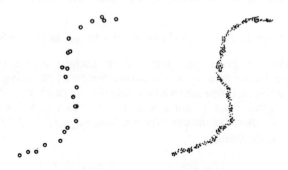

Figure 3.6. Points are replaced by clouds

A cloud is a set of random generated points grouped around a center x_0. "Load" is the number of points loaded into the cloud. The number of points and the size of the cloud together define its *density*. In the present example, the cloud has the shape of a parallelepiped spanned by the vectors v_1, v_2, v_3.

If a set of unstructured points is to be represented, one can draw a cloud around each point by taking v_1, v_2, v_3 orthonormal. The clouds is then three dimensional. If, however, a normal vector is known, we can outline the surface much better by taking $v_3 = 0$ and by letting v_1, v_2 be orthonormal vectors which are perpendicular to the normal vector of the surface. This draws a *flat* cloud approximately tangent to the surface. If one has structured data or some knowledge of tangent vectors, one can also set $v_1 = r_u$, $v_2 = r_v$, $v_3 = 0$ to obtain the same effect (see Section 5.1.1 for the computation of the tangent vectors r_u, r_v). In this case the load has to be proportional to the area of the cloud so that all clouds have the same density.

By modifying the random function (with some probability theory) one can get other shapes of clouds. Here is a ball cloud of radius R:

$$\text{Cloud}(x_0, \text{load}, v_1, v_2, v_3, R)$$
$$= \{\, x_i = x_0 + r \sqrt{1 - z^2} \cos q * v_1 + r \sqrt{1 - z^2} \sin q * v_2$$
$$+ r*z*v_3$$
$$|\, r = R * \sqrt[3]{\text{rand}()} \,;\, z = \text{rand}(); q = \text{rand}()*2*p \,, i = 1..\text{load} \,\}$$

The third root appearing in the formula yields a homogeneous filling-in of the ball. On the screen this will not appear homogeneous though, because of the projection. For special effects we may replace the third root by the square root; we may also replace the homogeneous ball filling by a homogeneous filling-in of the surface of the ball, etc.

3.4.3. Special Effects

3.4.3.1. Transparency

One of the new features of a cloud is that it has *density*. Since it is formed by a rather small number of points, there is a lot of space to look through. This yields a transparency, which can be controlled by the load parameter. In this sense a filled polygon, for instance, is a cloud with maximal density: all pixels of the occupied area are switched on, and no transparency occurs.

3.4.3.2. Contours

In contrast to brightness, a density is not invariant under projection. When a flat cloud is looked at orthogonally, then its *apparent density*, that is, the density of the image, coincides with the real density. If, however, the cloud is seen from the side, then its apparent density is increased because the points apparently sit on a smaller area. This gives a new picture symbol: *density to indicate direction*, as shown in Fig. 3.10 and 3.11 (see colour section).

3.4.3.3. Depth-Cuing

Any scalar stocked in the paint box such as the "z-value" for orthographic projection, can be used to change the appearance of the cloud. For instance distant clouds can be given smaller density. In Fig. 3.11 this has been used to add depth information.

3.4.3.4. Thickening

A flat cloud used for the representation of a structured discrete surface can be given some thickness by slightly displacing the points in the direction of the normal vector N. This is easily realized by taking v_3 (cf. Section 3.4.2) to be a small fraction of N instead of zero.

3.4.4. Space Curves

Space curves are usually drawn as curved *lines* on the screen, and in many cases, this representation is quite satisfactory. However, it contains no depth information. Depth-cuing may help indicate distant parts, but it cannot be used as an indicator for the local direction of a curve. A way to overcome this difficulty, is to replace the curve by a small tubular surface or by a band of polygons. In SFUMATO, the thickening of a parameterized curve, say c, can be obtained without using surfaces. For this we split up the curve in segments and then replace each segment by a *cylindrical cloud*. It can be obtained as follows. A moving frame $v_1(t)$, $v_2(t)$, $v_3(t)$ of orthonormal vectors is introduced along the curve, for instance such that $v_3(t)$ is the tangent vector of the curve at point $c(t)$, where t is the parameter of the curve.

The idea is shown in the figure. If we assume that $c(t)$ is the curve segment, with t ranging from t_0 to t_1, then the cloud looks as follows

$$\text{Cloud}(t_0, t_1, \text{load}, v_1, v_2, v_3, R)$$
$$= \{ \ x_i = c(T) + r * \cos q * v_1(T) + r * \sin q * v_2(T)$$
$$| \ T = t_0 + (t_1 - t_0) * \text{rand}() \ ; \ r = R * \sqrt{\text{rand}()} \ ; \ q = \text{rand}()*2*p \ , \ i = 1..\text{load} \}$$

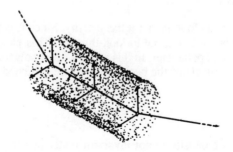

Figure 3.7. A cylindrical cloud around a curve segment.

A more refined method is to introduce $n(T) = \cos\theta\, v_1(T) + \sin\theta\, v_2(T)$ as a "normal vector" for each point x_i in the cloud and then to work with light sources. A special effect based on one moving vector $v_1(T)$ only, has been obtained in Fig. 3.12 (see colour section) with filled square clouds.

3.5. Differential Geometric Tools

In this section we review some definitions and facts of differential geometry which we used in the text. For an introduction to this subject we refer to (Nutbourne, 1988) or (Faux, 1979).

3.5.1. Normal Vectors

3.5.1.1. Parametric Surfaces

We first consider a parameterized surface S given by the equation

$$x = r(u, v), \quad (u, v) \in D.$$

The tangent plane of S at point x is spanned by the directional derivative vectors

$$r_u = \partial r(u, v)/\partial u, \quad r_v = \partial r(u, v)/\partial v.$$

The vectors r_u and r_v are also tangent to the parameter lines through x. The coordinates of r_u and r_v are computed by taking the partial derivatives of each coordinate $x_k = r_k(u, v)$, ($k = 1, 2, 3$).

The *direction* of the tangent plane is given by the *normal vector* $N(u, v)$ at $x = r(u, v)$, defined as

$$N = r_u \times r_v/|r_u \times r_v|,$$

where $\times$ is the cross product and $|\ |$ denotes the length of a vector.

3.5.1.2. Level Surfaces

For a level surface, given by the equation $f(x) - c = 0$, the normal vector $N(x)$ at x is given by

$$N = \text{grad } f(x) \,/\, |\text{grad } f(x)|$$

in which grad $f(x)$ is the *gradient vector*, $\text{grad } f = (\partial f/\partial x_1, \partial f/\partial x_2, \partial f/\partial x_3)$. The gradient vector is known to be orthogonal to the level surface.

3.5.1.3. Structured Discrete Surfaces

When a surface comes a priori in structured discrete form, then we cannot compute derivatives. Still a notion of *approximate* tangent vectors and normals can be maintained:

$$r_u[i][j] = r[i+1][j] - r[i][j], \;\; r_v[i][j] = r[i][j+1] - r[i][j]$$

and

$$N[i][j] = r_u[i][j] \times r_v[i][j]/|r_u[i][y] \times r_v[i][j]|.$$

3.5.2. Area

The area of a parallelogram spanned by the two vectors r_u, r_v
in space is given by $|r_u \times r_v|$.

3.5.3. Curvature

The interactive techniques of Section 3.1 support the visualization of the *global* properties of surface. There are also local features such as *curvature*.

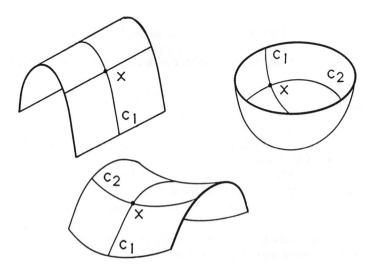

Figure 3.8. Surfaces with zero, negative and positive curvature

Fig. 3.8 shows three very different examples. The first example might have been obtained by folding a sheet of paper; it rolls out easily into the plane. The second example, oval shaped, tears up when we try to roll it out, and the third example, saddle shaped, crumbles.

This aspect of shape is captured in the notion of *curvature* and can be measured by analyzing how the normal vector changes its direction if we walk along different curves on the surface. One is thus lead to consider, at any point $\mathbf{x} = \mathbf{r}(u, v)$ of the surface, the directional derivative $\mathbf{N}_t$ of the normal vector:

$$\mathbf{N}_t = a\mathbf{N}_u + b\mathbf{N}_v$$

where $\mathbf{N}_u = \partial N(u, v)/\partial u$, $\mathbf{N}_v = \partial N(u, v)/\partial v$, and vector $\mathbf{t}$ is the linear combination $\mathbf{t} = a\mathbf{r}_u + b\mathbf{r}_v$. There are different kinds of curvatures: an observer walking across $\mathbf{x}$ in the direction of vector $\mathbf{t}$ feels the directional curvature

$$k(\mathbf{t}) = -\frac{<\mathbf{N}_t, \mathbf{t}>}{|\mathbf{t}|^2}.$$

The minimal possible value k_1 and the maximal possible value k_2 of $k(\mathbf{t})$ at $\mathbf{x}$ are called the *principal curvatures*. In Fig. 3.8 the principal curvatures are felt by an observer walking along curves c_1 (the minimal value) and c_2 (the maximal value). We refer to (Nutbourne, 1988, chapter 5) for a number of useful formulas about curvature, and also to (Faux, 1979, Section 3.4.2). Here we only mention the following. The principal curvatures are usually combined into the mean curvature H and the Gaussian curvature K. They are defined as follows.

$$H = \tfrac{1}{2}(k_1 + k_2), \quad K = k_1 k_2.$$

In Fig. 3.8 the Gaussian curvature is 0 in the first example, positive in the second and negative in the third.

The example of Fig. 3.2 and Fig 3.11 (see colour section) are surfaces of *constant* mean curvature, but this fact is not obvious from the figures. It is an interesting problem to find the appropriate picture symbols which visualize curvature properties of a surface.

References

Faux, I. D., and Pratt, M. J. (1979). *Computational Geometry for Design and Manufacture.* Ellis Horwood, Chichester.

Lucas, M. (1988). Elimination des parties cachées. In Liebling, Th., and Röthlisberger, H. (ed.), *Infographie et Applications.* Masson, Paris.

Magnenat-Thalmann, N., and Thalmann, D. (1987). *Image Synthesis.* Springer-Verlag, Tokyo.

Nutbourne, A. W., and Martin, R. R. (1988). *Differential Geometry Applied to Curve and Surface Design.* Ellis Horwood, Chichester.

4
Solid Modelling

N.F. Stewart*
Université de Montréal

This chapter is divided into five sections. In the first section we shall give an introduction, and an outline of the historical background of solid modelling, including a summary of the main representation methods available. In the second and third sections we shall discuss the underlying class of objects to be modelled, and criteria for comparing representation methods. In the fourth section, a study is made of the four principal kinds of methods: sweep methods, constructive methods (including Constructive Solid Geometry, or CSG), boundary representation methods, and decomposition methods. In the last section, a comparison of the four approaches, from the point of view of computing cost, memory requirements, and generality of representation, will be given, as well as a description of certain relationships amongst the four approaches.

4.1. Introduction and historical background

4.1.1. Origins of solid modelling

A retrospective look at the historical origins of Solid Modelling reveals three major avenues of relevant research: Computer Graphics, Computer-Aided Design and Manufacturing (CAD/CAM), and a fourth, much older, classical Geometry and Algebraic Topology. Computer Graphics had its beginnings in the 1950's with primitive techniques for the output of two dimensional drawings, and in the early 1960's with the first appearance of interactive graphic techniques. (A brief history of Computer Graphics from an American perspective is given in (Foley and Van Dam 1982)) This led at quite an early stage to methods for the representation of three dimensional objects, for example by wireframe diagrams, and to related questions such as hidden line elimination; wireframes not only permitted the graphical display of two dimensional entities, such as printed circuit boards, but also, graphical display of *objects*. Subsequently, (paraphrasing (Voelker et al. 1988)), polygonal (faceted, tiled) systems approximating the boundaries of solids with unstructured collections of planar polygons to produce visual effects for a variety of nonmanufacturing purposes (such as real-time imagery for flight simulators, commercial animation, and research in visual perception) were developed, as well as methods for the use of sculptured surfaces in the aero, marine, and

* This research was supported in part by a grant from the Natural Sciences and Engineering Research Council of Canada

automotive industries. By 1981 there was already a large literature in the area of Computer-Aided Geometric Design (Barsky 1981).

The graphical display of printed circuit boards, already mentioned, may be considered also to be part of CAD/CAM, and it represents a line of research which continues today with sophisticated systems for VLSI design and testing. But work in the area of CAD/CAM also progressed in the direction of automation of drafting systems, "electronic prototyping" (Hopcroft and Krafft 1887), and use of CAD/CAM in mechanical design. The first programming languages for numerically controlled (NC) machine tools, introduced in the 1950's (Voelker et al. 1988), and the introduction of electronic drafting, led naturally to the idea of linking the two components, so that the information defining *objects* could be sent directly to the machine tool (see however (Johnson 1986, p. II-17)). The applications envisaged for modern solid modellers in CAD/CAM include every "... step in the design of a mechanical product: layout and basic design; analysis; prototyping and testing; detailing and documentation; manufacturing engineering; and engineering management ..." (Johnson 1986, p. I-3).

Attention was drawn to the word "objects", in the two paragraphs above, by means of italics. It is today very clear that if we want to build reliable systems which permit graphical display of objects, and reliable systems which permit description of objects for tooling (and other) operations, then we should take care to define exactly what an object is. For example, it is well known that wireframe diagrams do not necessarily specify valid solids unambiguously: thus, in Figure 4.1, there may be a hole through the object in any of three possible directions. Similarly, the "unstructured collections of planar polygons" mentioned earlier need not specify a valid solid: the person or program defining the collection might inadvertently neglect to include one of the walls of a house, for example, or position a section of the roof so that it intersects the floor.

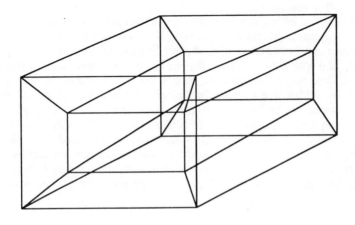

Figure 4.1. An ambiguous wireframe diagram

Up until the 1970's, almost no attention was given to the questions of ambiguity and invalidity in modelling solids. This was understandable when the field was new, since we cannot expect that all difficulties will be resolved immediately; furthermore, in the context of Computer Graphics, many applications (such as real-time imagery for flight simulators, commercial animation, and research in visual perception--see above) do not *necessarily* require valid solid models. In fact, sometimes in Computer Graphics one is only interested in generating interesting pictures, with no pretence that they model anything in the real world. On the other hand, it is less understandable in applications where we clearly intend to model three dimensional objects. Even today, not enough attention is given to these questions. It was still possible in 1989 to cause a certain commercial solid modelling system to fail by presenting it with the example in Figure 4.1; yet this is probably the most widely known example of ambiguity in Solid Modelling.

In a series of technical memoranda (including (Requicha 1978), which was seminal), Requicha and his collaborators laid the foundations for the careful study of what should be considered a valid object, and how such objects should be represented. In particular, it was cogently argued that this is a crucial question, because the automatic modelling systems of the future will not be able to rely on human intervention to resolve their internal anomalies. Requicha was led to make use of a great deal of classical Algebraic Topology in order to accomplish his goal. Researchers in solid modelling are not the first to ask themselves exactly what a reasonable model for an "object" might be: mathematicians have studied the question for quite a long time.

Solid modelling systems began to be built on an experimental basis in the 1960's and 1970's, and today there are many commercially available systems: two dozen of those available in 1986 were reviewed in (Johnson 1986). Many practical and theoretical problems remain, however, including those of numerical robustness, representation of non-rigid objects, the introduction of tolerances in the specification of objects, and the efficient representation of objects outside the class of *r-sets* (a term which will be defined below).

4.1.2. Physical objects, mathematical objects, and representations of objects

It is important to distinguish amongst the following, which are not the same: the physical object to be modelled, the mathematical object which models it, and the representation of the mathematical object (Requicha 1980). In this chapter, "object" and "solid" are synonyms, and refer to the mathematical object. Given a class of objects, there may be many different possibilities for their representation.

4.1.3. Principal methods of representation

The principal methods of representation may be divided into four classes: sweep methods, constructive methods (including Constructive Solid Geometry, abbreviated CSG); boundary representation methods; and decomposition methods, including octrees and Binary Space Partitioning (BSP) trees. Each of these major approaches will be discussed below.

4.2. The underlying modelling space

We shall restrict our attention to certain subsets of the class of r-sets, which are, intuitively, "...solid curved polyhedra bounded by well-behaved surfaces" (Requicha 1980). More formally, an *r-set* is a compact (closed and bounded) subset of Euclidean three space E^3 that is regular and semi-analytic. *Regular* means that the set is equal to the closure of its interior, which excludes pendent edges and faces. *Semi-analytic* means that the set can be described at any point by a finite set of real analytic functions (Lojasiewicz 1964, page 450). The precise meaning of the word "described" is given in the reference just cited, along with a theorem ensuring the existence of a *finite* triangulation of a semi-analytic set. The fact that any r-set admits a finite triangulation is of fundamental importance.

In the sequel we shall often refer to "regularized Boolean operations", which simply means that after effecting a Boolean operation (union, intersection, or set difference), the result is replaced by the closure of its interior, thereby producing a regular set as a final result. (In fact, we are interested exclusively in regularized operations here, and the adjective will soon be dropped.)

The choice of the basic class of mathematical objects to be used (that is, the underlying modelling space), is a subjective decision, and the statement that one or another class of objects is an appropriate choice is not capable of proof. Such a statement has, rather, the status of a *thesis*, which is justified only on the grounds that the class seems to exclude sets which clearly should be excluded, and includes the sets which clearly should be included. Perhaps the most famous example of such a thesis is *Church's thesis* (Kleene 1971, p. 300), which states that every effectively calculable function is general recursive. There is no proof that this statement is true: it is only that many people working in many different areas proposed definitions for what they believed should be called "calculable", and in the end, these definitions turned out to be equivalent.

An analogous (although perhaps less spectacular) situation obtains in our case. An examination of the literature in the decade following their original proposal (Requicha 1978) shows wide (explicit and *de facto*) acceptance of the r-sets as a modelling class. Indeed, the choice can be justified:

. by intuitive arguments about what is "manufacturable."

. on grounds of existence of a finite representation

. by closure properties with respect to regularized Boolean operations

. on grounds of sufficiency to support many standard representations proposed for modelling solids.

Of course, this does not exclude the possibility of other choices for the fundamental modelling space: different choices may be valid in different contexts. For example, it may turn out that for some purposes, even the concept of a rectifiable curve for the boundary of a region may be inappropriate (Mandelbrot 1983). Similarly, by choosing the r-sets for the class of underlying solids, we are restricting our attention to the case of nominal objects, supposed to be exactly defined, and involving no uncertainty in their definition: the more general case, where tolerancing information is added to the model, will lead to a new class of objects for the underlying modelling space (Requicha 1983). Yet another example is the case illustrated by two non-homogeneous r-sets glued together, so that there are internal faces: this too will lead to other choices for the modelling space. All of these generalizations are interesting, but for none of them do we yet have theoretical foundations as firm as those described above for the r-sets.

Another subset of the r-sets that has been widely used in the field of solid modelling is the class of *manifold objects*, which are r-sets for which every boundary point has a neighbourhood homemorphic to the unit (two-dimensional) disk (Mantyla 1988). One of the main advantages of this class of sets is that the Euler operators (described below) are available, and this is very natural for the construction and maintenance of objects when boundary representations are used. However, this class has the disadvantage that it is not closed under the regularized Boolean operations: for example, if we consider a flat slab resting on four pointed pyramids as a single object (the union of the four pyramids and the slab), we see that the class of manifold objects is not closed with respect to regularized set union. It is possible to extend the Euler operators to the r-sets in a natural way (Desaulniers and Stewart 1990), and therefore the r-sets seem to be a more appropriate choice for the underlying modelling space.

4.3. Criteria for comparing methods of representation

As stated above, given the underlying modelling space, there may be many choices for the method of *representing* a given object in the space, and there will normally be a trade-off involved, depending on which operations are to be performed most frequently. It may in fact be useful to maintain more than one representation, to provide different levels of abstraction (Requicha 1980, p. 29), or in order to use the representation most appropriate for each operation. (In this case, however, one of the representations should be chosen as the defining representation, and we must deal with the question of consistency amongst the representations used.) Some examples of the operations to be performed are:

- translational and rotational "sweeping" for purposes of input of objects
- regularized Boolean operations using existing objects as arguments
- sectioning of objects
- application of rigid motions to objects
- application of local modifications (rounding, blending) to objects
- implementation of an "undo" facility for any or all of the above operations
- parametric modification of a part or feature of an object
- graphical display of modelled objects
- detection of intersections between objects (interference checking)
- calculation of the distance between objects
- calculation of centre of mass, weight, volume or surface area of an object
- calculation of stresses in various parts of the object
- calculation of eddy currents in an object due to an enveloping magnetic field
- conversion to other representations
- generation of cutter paths for NC machine tools

In practice, a method of representation will only be capable of representing a subset of the set of r-sets (for example, it may permit only objects with non-curved boundaries). For most methods, this subset, like the set of r-sets itself, will be closed with respect to regularized Boolean operations, assuming infinite precision arithmetic. (This will not always be true, however: for example the intersection of two objects with boundaries defined by bicubic patches will not in general produce an object in the same class (Hopcroft and Krafft 1987), and if we have at our disposal only finite precision arithmetic, the finite number of floating point numbers in the machine will preclude the possibility of a strict closure property for many representation methods.) Closure with respect to rigid motions is similarly a desirable property of a representation method. With these remarks in mind, we can list criteria for choosing amongst different representations:

- *execution time* of algorithms for the operations defined above

- *memory requirements:* for the representation of objects, and for the algorithms for the operations defined above

- *geometric coverage*: how large a subset of the set of r-sets can be represented? If a sufficiently general form of the representation method is being onsidered, we might even ask about *completeness* of the representation: can all r-sets be represented? (This terminology does not correspond to (Requicha 1980, p. 9)—we prefer the alternate choice of "unambiguous" for the concept defined there. Our use of "completeness" is similar to that of (Mantyla 1984).)

- *precision*: if a given r-set, or the result of an operation such as a regularized Boolean operation or a rigid motion, is outside the subset of r-sets that a method can represent, what is the size of the error in the approximate result actually produced by the method?

- ease of ensuring *geometric validity* (this problem is always present, but may be more acute in the case hypothesized in the previous item). This is part of the more general question of *soundness* of the representation: can only r-sets be represented?

- *nonambiguity* and *uniqueness* of the representation.

Clearly, the answer to the question of what is the best method of representation may depend on the criterion used, and on the relative frequency with which various operations are to be performed.

4.4. Methods of Representation

What we shall refer to as the "usual" classification of representation methods (see, for example (MagnenatThalmann and Thalmann 1987, Ch.4; Samet 1990a, p. 317)) usually divides them into six classes. Of these, we shall not give "primitive instancing"—the *ad hoc* representation of (possibly parametrized) solids—the status of a class of representation methods. This leaves us with

- Sweep representations

- Constructive methods (including CSG)

. Boundary representations

. Decomposition methods (in which we combine two of the usual categories Magnenat-Thalmann and Thalmann 1987, Ch.4), spatial occupancy enumeration and cell decomposition).

Note that we have not included methods that are ambiguous, such as wire frame diagrams and (incomplete) surface models (Johnson 1986, p. II-5; Chiyokura 1988, p.6), although these may be useful for design purposes, and for topographic data (Samet 1990a, p. 365), and Miller (1989) argues that such models should be integrated into solid modelling systems.

The usual terminology for the various representation methods is not always a good guide to the essential differences amongst the methods. For example, representation of objects by means of their boundaries is an integral part of some constructive methods, and yet they are not included in the class of boundary representations. Some authors introduce other misleading terminology, which makes understanding difficult: for example, regularized Boolean operations are sometimes referred to as "CSG operations", because they appear explicitly in the tree used in the CSG approach. But Boolean operations (we shall henceforth omit the adjective 'regularized') can be and are applied to sets represented by all of the major representation methods. It is important to keep these different things clearly separated: we are looking for methods to represent elements of some subset of the modelling space (here, the r-sets). The Boolean operations are just that: *operations* on sets in the modelling space; these operations must be supported by any representation method.

Most representation methods are based on some sort of finite graph structure, and usually (except for boundary representation methods) on some sort of tree. The structure of the tree, and the information stored at the nodes of the tree, will vary depending on the method.

4.4.1. Sweep representations

Sweep representations may be translational sweeps, rotational sweeps, or more general sweeps. To represent an object by a translational sweep it is sufficient to give two distinct parallel planes, a two dimensional region in one of the planes, and a vector (direction of sweep) not parallel to the planes. The object is defined by moving the two dimensional region, along the direction of sweep, from one plane to the other. A rotational sweep is similar, except that the region is moved around an axis of rotation. More general kinds of sweep representations can also be defined (Requicha 1980, p. 57; Magnenat-Thalmann and Thalmann 1986, Sec. 4.4.3.4). As indicated above, in Section 4.3, sweeping is often used as a method of input of objects, independently of its use for the purpose of object representation.

4.4.2. Constructive methods

Since the r-sets are described at each point by a finite number of analytic functions, it is natural to take as a primitive object an analytic half-space $\{x : f(x) \leq 0\}$, where f is a real analytic function, and to permit any finite Boolean combination (union, intersection, set difference) of such half spaces. This approach to representation is called a *half-space model*. For example (Mantyla 1988, p. 78), we might define a finite cylinder as the intersection of the three primitive halfspaces $x_1^2 + x_2^2 - r^2 \leq 0$, $-x_3 \leq 0$, and $x_3 - h \leq 0$.

There is more than one choice for the actual representation of half-space models: Boolean combinations may be brought into "sum of products" form (Mantyla 1988, p. 80), or the hierarchy of Boolean operations to be formed can be represented explicitly in a tree, with one Boolean operation at each node in the tree. The latter approach is central to the CSG method, to be discussed presently. The thing that distinguishes the half-space method as a constructive method is that *arbitrary finite Boolean combinations can be explicitly specified and retained for later evaluation*. Later we shall discuss the BSP-tree approach, which is based on a tree structure, and which also defines sets in terms of half-spaces subject to certain Boolean operations. The fundamental difference is that the structure of BSP-trees implicitly determines the Boolean operations applied: arbitrary sequences of Boolean operations will not be specified in the tree.

Note also that the primitive sets used here (the half-spaces) are defined explicitly in terms of their boundary, which illustrates that whether explicit boundary representations are used is not a reliable way to place a method with respect to the usual classification of representation methods.

A second example of a constructive modelling system, and one of the major approaches to representation, is *Constructive Solid Modelling*, or *CSG*. The difference between this approach and the half-space models just described is that a fixed set of compact primitive objects is used: it might include, for example, finite cylinders, prisms, hyper-rectangles, and so on (the exact list depending on the particular system). The set of interest is represented as a finite tree (or, if subtrees are to be shared, by a directed acyclic graph), with Boolean operations, and/or rigid motions (rotations and translations), at the nodes of the tree. Again, the defining characteristic of a constructive system is that arbitrary Boolean operations may be explicitly specified for later evaluation. (Another possible use for specifying the Boolean combinations involved in the construction of an object, even if they are performed immediately, is to keep a record of what has been done, in order to be able to undo the sequence of operations. This approach is actually implemented in the context of a boundary representation in (Chiyokura 1988); indeed, since the hierarchy of operations recorded in the tree includes all Boolean operations, the system described there could be considered a type of CSG system, with primitive objects defined by a boundary representation.) The method of representing the primitive objects is completely arbitrary. The primitive objects could in principle be represented in any of several different ways: an *ad hoc* approach permitting definition of a small number of parametrized solids (for example, representation of a sphere by specifying its centre and its radius), or alternatively, by any of the methods to be described below (a simple boundary representation, a BSP-tree, or other method). In the latter case, most authors would say we are dealing with a "hybrid" method.

4.4.3. Boundary Representations

Rather than represent sets as finite Boolean combinations of primitive point sets whose method of representation may be invisible to the user of the system, one may try to represent sets by specifying their boundary explicitly. The mathematical question of whether a set *can* be specified by specifying its boundary is an old and interesting mathematical question (Requicha 1978) which was answered negatively by Brouwer at the beginning of the century (Alexandroff 1961, p.18). However, in the special case of the r-sets, a verifiable condition for a set of boundary "patches" to define a unique r-set is available (Requicha 1978, Section 5.2). It therefore makes sense to study methods for representing r-sets by specifying their boundaries.

The *boundary representations* described here differ from the constructive representations of Section 4.4.2 in that there is no provision here for the specification of Boolean operations to be performed at some later time. Objects are stored as explicitly defined entities. It is usually said that they are stored in *evaluated* form, in contrast to the *unevaluated* objects implicitly defined by the operations contained in a CSG tree: the boundary of the object may be "evaluated" for purposes such as graphical display. Use of the boundary to represent an object is very natural in Computer Graphics, given the emphasis on display of objects; in fact, as mentioned in Section 4.1.1, representation of objects by unstructured collections of planar polygons was used quite early. For many applications in Computer Graphics, this approach may be sufficient.

A data structure which permitted specification of the relationship amongst the edges, vertices and faces defining the boundary was introduced by Baumgart (Baumgart 1974): the *winged-edge data structure*. This data structure and its derivatives are still widely used today. Alternative boundary data structures are described in (Ansaldi et al. 1985; Weiler 1985; Woo 1985).

The advantage of maintaining such explicit information about the relationship amongst vertices, edges, and faces of the boundary of the (perhaps curved) polyhedron is that it facilitates maintenance of (at least) *topological validity* at each step, when manipulating objects for sectioning operations, Boolean operations, and so on, since there is a theory of realizability that corresponds very closely to the entities used in such a representation. It is advantageous to ensure such validity at each step (if possible) because *a posteriori* validity checking is very expensive. The theory is based on the Euler-Poincaré equation, which for a polyhedron (homeomorphic to a sphere) becomes $F - E + V = 2$, where F is the number of faces in the polyhedron, E is the number of edges, and V is the number of vertices. Baumgart introduced a set of operators, called the *Euler operators*, which can be used in the construction and manipulation of object boundaries (Chiyokura 1988, p. 18). Application of an Euler operator maintains the correctness of the Euler-Poincaré equation. For example, one of the operators is MEV ("make edge and vertex"), which adds an edge and a vertex to the topological structure defining the boundary, and thereby maintains the value of $F - E + V$. The inverse operator is KEV ("kill edge and vertex"), which also leaves $F - E + V$ unchanged. An easily accessible reference for the Euler operators is (Mantyla 1988), where they are used as basic primitives, so that one is sure after any operation that at least *there exists* a realizable three dimensional object corresponding to the graph defined by the winged edge structure.

The Euler-Poincaré relation (in its more general form) is the basis for a theory of physical realizability for two dimensional *closed manifolds*, that is, for objects whose boundaries are homeomorphic to a finite number of spheres with handles (like the surface of a many handled coffee cup) (Mantyla 1984; Mantyla 1983). If we require the boundary of the object to be a closed manifold, and add the restriction that the physical realization should be an r-set, then we have an alternative candidate for the underlying modelling space, the *manifold solids*. The disadvantage of this choice is that the class of manifold solids is not closed under Boolean operations, regularized or not: the example of the slab resting on four pointed pyramids, mentioned in Section 4.2, shows for example that the manifold solids are not closed under the union operation. (It can be shown that regularized union is the same as the union operation. No such result holds for intersection and set difference.) But of course, it is important to implement Boolean operations for objects represented using boundary representations (Mantyla 1988), and this is a very unsatisfying incompatibility.

The advantage of the manifold solids is that the theory of Euler operators is available for the manipulation of object boundaries. However, as was also mentioned in Section 4.2, it is possible to extend the Euler operators to the r-sets in a natural way, and to introduce a corresponding boundary data structure for the r-sets (Desaulniers and Stewart 1990). (See also (Laidlaw et al. 1986)) This appears to eliminate any trade-off between the manifold solids and the r-sets as a choice for the underlying modelling space: all of the advantages lie with the r-sets.

One of the most active areas of research in the boundary representation area has been the use of free-form surfaces, such as B-splines and Beta-splines, for the definition of boundaries (Chiyokura 1988, Ch. 5-8; Farin 1987; Mortenson 1985).

4.4.4 Decomposition methods

The central idea behind the *decomposition* methods is to decompose E^3 into subsets, and to indicate by some means which subsets are part of the object to be represented, and which parts are not. For both of the methods presented here, the structure of the decomposition of Euclidean space is defined by a tree.

Included in the class of decomposition methods, but not discussed here, are the *cell decomposition* methods (Requicha 1980, p. 39) (Mantyla 1988) and the related *Finite Element Methods* used in Engineering and in Computer Graphics, which are discussed in Chapter 5.

The decomposition methods, and the boundary representations discussed in Section 4.4.3, have in common that they do not permit explicit definition of an arbitrary hierarchical set of Boolean and rigid motion operations on primitive objects. They are also similar in that the more elaborate implementations of one decomposition method (the BSP-tree) in fact incorporate some boundary information. However, this boundary information is redundant: the basic approach, to specifying which points form part of an object, is quite different.

The decomposition methods discussed here are like the constructive methods in that they are based on a tree structure which implicitly defines a sequence of operations on certain subsets of E^3, and so in some sense they are "unevaluated" representations. But the decomposition methods are different from constructive methods in their basic approach to specification of which points form part of an object, and as we have noted, they do not permit explicit definition of an arbitrary hierarchical set of Boolean operations.

Two important examples of decomposition methods are BSP-trees and octrees. Although octrees were introduced as a representation method before BSP-trees, we shall first present the latter method, since the former may be viewed as a special case.

BSP-trees were first introduced as a method for solving the hidden surface problem in the display of collections of faces, perhaps associated with a boundary representation of an object, in contexts where the environment tends to remain static, and only the viewer's position changes (see (Fuchs et al. 1980), where the original idea is attributed to (Shumaker et al. 1969)); they are also of interest in ray-tracing (Dadoun et al. 1985). It was later proposed (Thibault and Naylor 1987) to use them as an independent method of representation of objects.

We begin by defining a *generic* BSP tree (Thibault and Naylor 1987). Such a tree is a *full* binary tree (Standish 1980, p. 51). Corresponding to each internal node is an open region, and a (sub)-hyperplane which divides the region into two disjoint non-empty subregions; these two open regions are associated with the children of the node. Corresponding to each leaf is a region, and a label ("in" or "out") which specifies whether the region forms part of the object or not. The region corresponding to the root node is E^3 itself. A simple example is given in Figure 4.2.

One could envisage generalizing this idea by replacing the (linear) hyperplanes dividing the region by "analytic hyperplanes" of the form $\{x : f(x) \leq 0\}$, although the original motivation for introducing the BSP-trees would then be lost, since the hidden surface algorithm (Fuchs et al. 1980) depends crucially on the linearity of the hyperplanes. It is interesting to note, however, that we have come full circle: this generalized form of the BSP-trees may be viewed as a tree representation of the "sum of products" form of the half-space model described in Section 4.4.2.

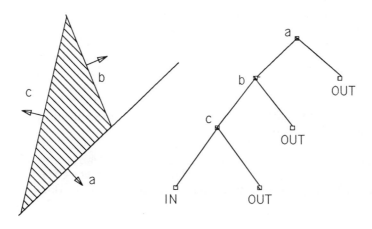

Figure 4.2. An example of a generic BSP-tree

A generalization of the *generic* BSP-tree introduced above is the *augmented* BSP-tree (Thibault and Naylor 1987), in which the nodes may have polygonal sets (representing faces of the object) associated with them. In the generic tree, each node corresponds to a convex region of the object. In a Computer Graphics application, it may be of interest to associate one or more coloured faces with a side of the convex region. For example, if the side represents the side of a building, there may be a face corresponding to each window in the building, where the windows are coloured differently from the building itself, and can be distinguished when the object is rendered on a graphic screen. The supplementary faces in an augmented BSP-tree may be viewed as defining boundary information, but the faces are not linked together as they were in the boundary representation, and so we view the BSP-tree as a decomposition method containing supplementary boundary information. (Of course, one could, if one wished, view the augmented BSP-tree as a boundary representation for which the logical relationships amongst the faces are given not by pointers linking vertices, edges and faces, as in the winged-edge structure described in Section 4.4.3, but rather by a binary tree structure defining the relative positions of the faces in space. Like any classification of methods, this one is to some extent arbitrary.)   Some related data structures are described in (Samet 1990a) and (Mehlhorn 1984, Vol.3-VIII-Ch.5).   Suppose now that the root of the binary tree defining (for simplicity, a generic) BSP-tree is said to be at level 1 (Standish 1980, p. 48), its children at level 2, and so on. If the leaves of the tree appear only when the level i is equal to 1 (*modulo 3*), if the normals to the internal node hyperplanes are $[0,0,1]$, $[0,1,0]$ or $[1,0,0]$, depending on whether the level i of the node is equal to 0, 1 or 2 (*modulo 3*), and if the tree satisfies the condition that both children of an internal node at level i equal to 1 or 2 (*modulo 3*)   must themselves be internal, then we get a tree structure that is equivalent to an *octree*.   The octree divides E^3 into isothetic hypercubes whose leaves are labelled "black", "white", or "grey" (corresponding to leaves marked "in", leaves marked "out", and internal nodes, respectively, in the language of BSP-trees.) There is a large literature on the topic of octrees (Samet 1990a; Samet and Webber 1988; Samet 1990b).

4.5. Differences and similarities amongst the principal methods

Given the large number of criteria for comparing methods, and the large number of different kinds of operations which may be of interest, as described in Section 4.3, it will only be possible to compare the principal approaches in very general terms. The task is complicated by the fact that the use of more than a single representation, and the use of hybrid methods, is very common. We separated the basic approaches in order to clarify the ideas involved, but in practice, various combinations may be useful. A good example is given by (Thibault and Naylor 1987), which was cited above as a paper dealing with (augmented) BSP-trees. In fact, however, (Thibault and Naylor 1987) discusses the application of Boolean operations to polyhedral objects defined by boundary information (and perhaps by an explicit boundary representation). In the context of the above discussion, this should be viewed as a CSG representation with primitive objects defined by a (more or less sophisticated) boundary structure.   (Again, most authors would call this a "hybrid" CSG and boundary representation.) The authors of (Thibault and Naylor 1987) think in terms of "evaluation" of the results of the Boolean operations involved in the definition of a CSG object, for purposes of graphical rendering; this is done by converting the result of the Boolean operations to a BSP-tree.

4.5.1. Comparison of the principal approaches

Considering first the criterion of *execution time* of algorithms, the CSG approach formally admits a very inexpensive algorithm for computing Boolean operations and rigid motions, using existing objects as arguments, since the arguments are simply inserted into a larger CSG tree specifying the operation. Of course, if the boundary of the object must be "evaluated", we have only postponed the problem; however, it may be possible to render A o B, where o is a Boolean operation, without actually computing A o B (Samet 1990a, p. 339; Sato et al. 1985; Goldfeather et al. 1989). On the other hand, if ray-tracing (Mantyla 1988, p. 92; Glassner 1989) must be used, the cost is quite high, although the use of hybrid representations, like those described above (Thibault and Naylor 1987), and parallel hardware for rendering, can greatly reduce this cost.

The cost of computing Boolean operations for objects represented by boundary representations is quite high (Mantyla 1988). Calculation of Boolean operations for octrees is surprisingly straightforward (Samet and Tamminen 1983); on the other hand, even a simple rotation is difficult (aside from problems of precision, discussed below). See also (Mantyla 1988, p. 71; Samet 1990b).

Local operations cannot be as conveniently implemented for CSG representations as they can for boundary representations (Johnson 1986, p. III-24; Chiyokura 1988, p. 22, p. 151). Intersection and distance calculations are discussed elsewhere in Chapter 17. Conversion algorithms are discussed in (Requicha 1980; MagnenatThalmann and Thalmann 1986; Samet 1990a). Sweeping operations are discussed for example in (Chiyokura 1988; Requicha 1980b).

As for *memory requirements*, boundary representations tend to require much larger quantities of memory than CSG representations. This is especially true if curved objects are approximated by (linear) polyhedral boundaries (Mantyla 1988, p. 119).

The *geometric coverage* of CSG and boundary representations depends on the generality of the version implemented. It is common, for example, to restrict boundary representations to polyhedral faces, and to restrict CSG methods to a fairly small number of primitive objects. Restriction of boundary modelling systems to manifold objects is a serious limitation, because of the lack of a closure property for the Boolean operations; however, it is probably more desirable than introduction of *ad hoc* solutions to append non-manifold objects, when they arise, to a manifold object data structure. The geometric coverage of BSP trees is as good as a boundary representation with polyhedral faces, but the special case of octrees restricts the class of representable objects to unions of (isothetic) cubes. This means that a large number of cubes may be required to obtain reasonable accuracy.

CSG representations are trivially closed with respect to Boolean operations and rigid motions, but again, if boundary evaluation is necessary, for example, for graphical rendering, the difficulty is only postponed. Boundary representations are not closed with respect to Boolean operations if curved boundary segments are permitted, and finding efficient algorithms to compute *precise* representable approximations is an interesting problem (Chiyokura 1988, p. 19; Mortenson 1985, Ch. 7). As mentioned above, octrees are not even closed with respect to rotations (and for this reason, it was suggested in (Major et al. 1989) to associate a matrix defining a rigid motion with each octree, in order to avoid both the computational cost of rigid motions, and the imprecision introduced by successive rotations of an object represented by an octree). CSG representations, boundary representations and BSP-trees (including octrees) are *sound, unambiguous,* but not *unique* (except that there is a unique octree for those r-sets that are representable by an octree).

Assurance of *geometric validity* is a difficult problem for boundary representations, although the work in (Mantyla 1984) on *topological validity* at least permits us to ensure that *there exists* a realizable solid corresponding to the graph structure specified in the boundary model. If finite precision arithmetic is used, strict closure no longer holds for most operations. Precision and validity in this case are studied under the heading of "numerical robustness", which is currently a very active area in Computational Geometry (Hoffmann 1989). An overall comparison of boundary and CSG representations from the user's point of view is given in (Johnson 1986, p. III-23).

4.5.2. Relationships amongst the principal approaches

We have mentioned many relationships amongst the principal representation methods, and we conclude by recalling some of them.

First of all, the octrees may be viewed as special cases of BSP-trees. Some authors have suggested associating supplementary faces with the cells of an octree, in order to avoid the problem of poor approximation of curved surfaces by isothetic hypercubes (Yerry-Shephard83).

For different reasons, mentioned above, it has also been proposed to associate face information with the nodes of a BSP-tree. Thus, these methods can be viewed as decomposition methods with supplementary information about the boundary, or as boundary methods with an unconventional method of linking together the faces.

Decomposition techniques are also used as geometric directories in conjunction with other methods, as mentioned for BSP-trees in Section 4.4. See also (Mantyla 1988, p. III-23; Mehlhorn84).

It has been shown (Sederberg et al. 1985) that the surface patches proposed for use in boundary representations with curved surfaces normally correspond to algebraic surfaces (perhaps of very high degree), and these surface patches therefore normally specify half-spaces. (See however (Hopcroft and Krafft 1987, p. 15).) This shows a close relationship between the constructive and boundary representations.

Because virtually any class of object can be used for the set of primitive objects in a CSG representation, and at the same time, systems that use other representations may at least temporarily store a list of Boolean and/or rigid motion operations (see for example (Johnson 1986, p. III-24; Mantyla 1988, p. 119; Chiyokura 1988)), the distinction between CSG systems and other systems is often difficult to discern. Indeed, given a CSG system with a wide range of primitive objects, defined for example by a boundary representation, it is not clear how it differs logically from a system with parametrized solids as primitive objects, where the parameters can be specified by the user. In both cases, the user has some (perhaps limited) control over the choice of primitive objects. Since it is becoming so common to see CSG representations based on generalized CSG primitives}, thus creating methods which are hybrid (Mantyla 1988, p. 96) with respect to the usual classification of methods, it is advantageous to modify the standard classification. The important point to note is that is is reasonable to superimpose CSG operations on top of almost any other representation.

All of the methods in the standard classification except CSG share the property that they represent objects as a *disjoint* (with respect to regularized intersection) union of sets, that is, as the *direct sum* of disjoint sets. The tree defining Boolean operations and/or rigid motions ("the CSG-tree") can be added on top of any of these other methods, and in practice, has in fact been used on top of most of them. The use of the CSG-tree then becomes a question of

   . the timing of the actual operations (It may be decided to postpone evaluation of certain operations, for example, for reasons of space, and to re-evaluate them each time they are needed. The operations will in this case be recorded in the CSG-tree.)

   . the omission of the actual operations (Evaluation of the operations may in fact be postponed indefinitely. For example, it may be possible to render $A \cup B$ or other

Boolean operations without actually computing the operation (Samet 1990a, p.339; Sato et al. 1985; Goldfeather et al. 1989), to compute

$$\text{dist}(\mathop{U}_{i=1}^{m} A_i, \mathop{U}_{j=1}^{n} B_j)$$

without actually computing the unions (Hurteau and Stewart 1990), or to compute mass properties without computing the Boolean operations (Miller 1988, p.41). Similarly, in the context of collision detection (see Chapter 17} for example, it may be possible and efficient to detect that $A \cap B$ is null by enclosing A and B in bounding sets of some kind and avoiding the calculation of the actual intersection.)

. permitting the possibility of undoing operations (Chiyokura 1988).

. the choice of primary or archival representation. For example, to avoid cumulative error, and/or for reasons of space, it may be decided to use the unevaluated CSG-tree as the archival form.

In addition to the decisions implied by this list, there are decisions to be made concerning which (that is, evaluated or unevaluated) of the representations the user and/or applications developer has access to, and when (Johnson 1986, p. III-24; Mantyla 1988, p.126; Miller 1989, Figs. 10a, 10b).

We refer to any method that represents some sub-class of the r-sets as the union of disjoint (with respect to regularized intersection) subsets of E^3, as a *direct sum* method. This class includes all of the standard methods except CSG. One possible classification of these methods is shown in Figure 4.3. There, a method which is a special case of another is shown as a descendant. (For Bintrees, octrees, and the single half-space, it is necessary to restrict attention to some compact subset of E^3} if we want to obtain only r-sets.)

Many of the methods are capable of representing only a sub-class of the r-sets. For example, the boundary representations typically are restricted to the representation of those r-sets with boundary patches which are polygons, or which are chosen from some restricted class of curved surfaces. Cell decompositions and primitive instancing (which is a cell decomposition comprising a single cell) normally have the form of their cells restricted to some small parametrized class of objects. BSP-trees are usually restricted to (linear) polyhedral objects, and octrees are even more restricted: only isothetic cells are used in the representation. There is one method in Figure 4.3 which is not normally given the status of a representation method, since it is capable of representing only one type of set, *viz* a single half-space. This method is *only* useful when a CSG-tree is added (in which case it becomes the usual half-space method). Given any direct sum method m, we may decide for one reason or another to superimpose a CSG-tree, and obtain thereby a method which will be denoted CSG/m:

. CSG/(single half-space) is the standard half-space method (as we stated just above).

. When we add a CSG-tree to primitive instancing we get CSG/(primitive instancing), which is standard CSG.

. Both CSG/(Sweep) and CSG/(Boundary representation) are quite reasonable choices (Mantyla 1988, p. 96).

. Both CSG/(Boundary representation) and CSG/(BSP-tree) are reasonable (Thibault and Naylor 1987; Goldfeather et al. 1989; Cajolet 1989).

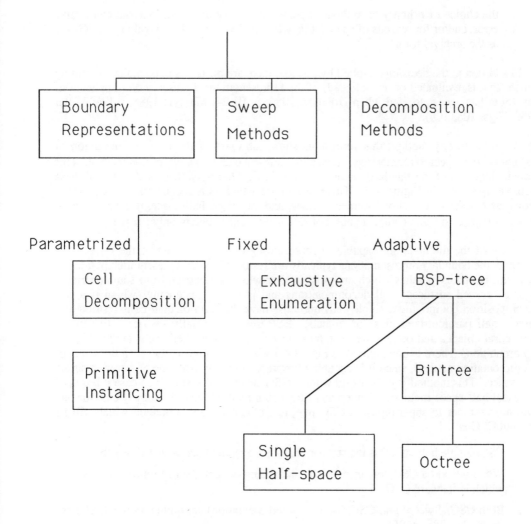

Figure 4.3. Classification of Direct Sum methods

References

Alexandroff, P. *Elementary Concepts of Topology*. Dover, 1961.

Ansaldi, S., De Floriani, L. and Falcidieno, B. Geometric modeling of solid objects by using a face adjacency graph representation. *SIGGRAPH* 131-139, 1985.

Barsky, B. Computer-aided geometric design: A bibliography with keywords and classified index. *IEEE C. G. and A.* 67-109, July 1981.

Baumgart, B. Geometric modelling for computer vision. PhD Thesis, Stanford University (also as Tech. Report CS-463), 1974.

Cajolet, C. Les opérations booléennes et les arbres binaires de partition de l'espace. MSc thesis, Université de Montréal, 1989.

Chiyokura, H. *Solid Modelling with Designbase*. Addison-Wesley, 1988.

Dadoun, N., Kirkpatrick, D. G. and Walsh, J. P. The geometry of beam tracing. Proceedings of ACM Symposium on Computational Geometry 55-61, 1985.

Desaulniers, H. and Stewart, N. F. *An extension of manifold boundary operations to the r-sets*. Manuscript, Dép't I. R. O., U. de Montréal. Submitted for publication.

Farin, G. E. (ed). *Geometric Modelling: Algorithms and New Trends*. SIAM 1987.

Foley, J. D. and Van Dam, A. *Fundamentals of Interactive Computer Graphics*. Addison-Wesley, 1982.

Fuchs, H., Kedem, Z. M., and Naylor, B. On visible surface generation by a priori tree structures. *SIGGRAPH* 124-133, 1980.

Glassner, A. ed. *Ray Tracing*. Academic Press, 1989.

Goldfeather, J. et al. Near real-time CSG rendering using tree normalization and geometric pruning. *IEEE C. G. and A.* 20-28, May 1989.

Hoffmann89 Hoffmann, C. M. The problems of accuracy and robustness in geometric computation. *IEEE COMPUTER* 31-41, March 1989.

Hopcroft, J. E. and Krafft, D. B. The Challenge of Robotics for Computer Science, in *Advances in Robotics*, J. T. Schwartz and C.-K. Yap, eds. Lawrence Erlbaum Associates, 1987.

Hurteau, G. and Stewart, N. F. Collision prevention in robotic simulation. To appear, *INFOR*, 1990.

Johnson, R. H. *Solid Modeling*. North-Holland, 1986.

Kleene, S. C. *Introduction to Meta-Mathematics*. North-Holland, 1971.

Laidlaw, D. H., Trumbore, W. B., and Hughes, J. F. Constructive solid geometry for polyhedral objects. *SIGGRAPH* 161-170, 1986.

Lojasiewicz, S. Triangulation of semi-analytic sets. *Annali della Scuola Normale Superiore de Pisa, Annali Scienze, Fisiche i Mathimatiche Serie* III(18) 449-474, 1964.

Magnenat-Thalmann, N. and Thalmann, D. *Image Synthesis: Theory and Practice*. Springer-Verlag, 1987.

Major, F., Malenfant, J. and Stewart, N. F. Distance between objects represented by octrees defined in different coordinate systems. To appear, *Computers and Graphics*, 1990.

Mandelbrot, B. B. *The Fractal Geometry of Nature*. Freeman and Company, 1983.

Mantyla, M. A note on the modeling space of Euler operators. *Computer Vision, Graphics and Image Processing* 26, 45-60, 1984.

Mantyla, M. Computational Topology. *Acta Polytechnica Scandinavica*, Mathematics and Computer Science Series No. 37, Helsinki, 1983.

Mantyla, M. *Solid Modeling*. Computer Science Press, 1988.

Mehlhorn, K. *Multi-dimensional Searching and Computational Geometry*. Springer-Verlag, 1984.

Miller, J. R. Analysis of quadric-surface-based solid models. *IEEE C. G. and A.* 28-42, January, 1988.

Miller, J. R. Architectural issues in solid modelers. *IEEE C. G. and A.* 72-87, Sept. 1989.

Mortenson, M. E. *Geometric Modeling*. Wiley, 1985.

Requicha, A. A. G. *Computer Aided Design, Modelling, Systems Engineering, CAD-Systems*. Lecture Notes in Computer Science 89, J. Encarnacao, ed. Springer-Verlag, 1980.

Requicha, A. A. G. *Mathematical models of rigid solid objects*. Technical Memo TM-28, University of Rochester, 1977.

Requicha, A. A. G. Representations for rigid solids: theory, methods and systems. *Computing Surveys* 12, No. 4, 437-464, 1980.

Requicha, A. A. G. Toward a theory of geometric tolerancing. *Int. J. Robotics Research* 2, No. 4, 45-60, 1983.

Samet, H. and Tamminen, M. Computing geometric properties of images represented by linear quadtrees. *IEEE P. A. M. I.* 7(2) 229-240, 1985.

Samet, H. and Webber, R. Hierarchical data structures and algorithms for computer graphics. *IEEE C. G. and A.* 48-68, May 1988 and 59-75, July 1988.

Samet, H. *Applications of Spatial Data Structures*. Addison-Wesley, 1990.

Samet, H. *The Design and Analysis of Spatial Data Structures*. Addison-Wesley, 19 1990.

Sato, H. et al}. Fast image generation of constructive solid geometry using a cellular array processor. *SIGGRAPH* 95-102, 1985.

Schumaker, R. A. et al. Study for Applying Computer-Generated Images to Visual Simulation. AFHRL-TR-69-14, U. S. Air Force Human Resources Lab 1969.

Sederberg, T. W., Anderson, D. C. and Goldman, R. N. Implicit representation of parametric curves and surfaces. *Computer Vision, Graphics and Image Processing* 28, 72-84, 1984.

Standish, T. A. *Data Structure Techniques*. Addison Wesley, 1980.

Thibault, W. C. and Naylor, B. F. Set operations on polyhedra using binary space partioning trees. *SIGGRAPH* 153-162, 1987.

Voelcker, H. B., Requicha, A. A. G., and Conway, R. W. Computer Applications in Manufacturing. *Ann. Rev. Comput. Sci.* 3, 349-387, 1988.

Weiler, K. Edge-based data structures for solid modeling in curved-surface environments. *IEEE C. G. and A.* 21-40, January 1985.

Woo, T. A combinatorial analysis of boundary data structure schemata. *IEEE C. G. and A.* 19-27, March 1985.

Yerry, M. A. and Shephard, M. S. A modified quadtree approach to finite element mesh generation. *IEEE C. G. and A.* 39-46, January-February 1983.

5

The Finite Element Method: Basic Principles

Jean-Paul Gourret
Ecole Nationale Supérieure de Physique de Marseille

5.1. Introduction

The Finite Element Method (FEM) is fifty years old and has benefited from a great deal of refinement since its first appearance in structural engineering for solving the Boundary Value Problems (BVP) of elasticity. During these years, it appeared in various scientific and technical fields such as: fluid mechanics, aeromechanics, biomechanics, geomechanics, thermal analysis, acoustic, electric and magnetic field modeling, shock and impact modeling, reactor physics, plasma modeling ... allowing facilities for modeling coupled systems.
In these fields, the method was developed directly from analogies between different disciplines (Ratnajeevan and Hoole 1989) or from specific problems.

A presentation of the finite element method in a few pages is not an easy task. As the field is very large, we will give in this chapter general ideas about the FEM, taking examples from solid mechanics. Our purpose is to supply sufficient basic information, vocabulary and bibliography on the subject to readers not familiar with the method.

In section 2 we describe the integral formulation used for the discretization of a Boundary Value Problem.

In section 3 we describe the numerical calculations starting from the integral formulation, and up to the matrix solution for a linear continuum mechanics example.

The last section is an introduction to non-linear and dynamic analysis.

5.2. Boundary value problem and integral formulation

The state of a physical system may be described by a set of state variables $\mathbf{u} = \mathbf{u}(x,t)$ where x is the space variable and t is the independent time variable. The physical system is discrete when the number of degrees of freedom or parameters needed to define $\mathbf{u}$ is finite, and it is continuous when the number of degrees of freedom is infinite. The behavior of the system may be expressed by a set of algebraic equations or by partial differential equations, depending on the discrete or continuous nature of the system, associated with boundary conditions in space and in time.

In general terms, the state variables **u** must satisfy a set of linear, quasi-linear or non-linear differential equations in Ω, an open subset of R^N.

$$A(u) = L(u) + f_w = 0 \tag{5.1}$$

and a set of boundary conditions on $\Gamma = \partial\Omega$

$$B(u) = 0 \tag{5.2}$$

Boundary conditions are mainly linear, of Dirichlet, Neumann or Cauchy type, and may be homogeneous or not.

$$u|_\Gamma = g_0 \qquad (5.3) \qquad \text{Dirichlet essential boundary condition,}$$

$$\left.\frac{\partial u}{\partial n}\right|_\Gamma = g_1 \qquad (5.4) \qquad \text{Neumann natural boundary condition specifying the flux,}$$

$$\alpha_1 u|_\Gamma + \alpha_2 \left.\frac{\partial u}{\partial n}\right|_\Gamma = g_2 \qquad (5.5) \qquad \text{mixed Cauchy boundary condition specifying a combination of the flux and the state variable.}$$

the components of **u** may be scalars or vectors, **n** the outer unit normal vector to $\partial\Omega$, and **A** a single or multiple linear or non-linear operator.

Dropping the most general non-linear operators (Courant and Hilbert 1966), the normal form for a linear second order differential operator which includes a large number of physical systems is

$$L(u) = -\sum D_j(a_{ij}D_i u) + \sum b_i D_i u - \sum D_i(c_i u) + d\,u \quad \text{for } 1 \pounds i, j \pounds N \tag{5.6}$$

where $D_i = \dfrac{\partial}{\partial x_i}$, and a_{ij}, b_i, c_i, d are continuously differentiable and not simultaneously vanishing functions of **x** in Ω.

Boundary Value Problems for partial differential equations belong to elliptic, parabolic or hyperbolic types depending on the coefficients in (5.6). Parabolic and hyperbolic types are also called initial value problems because they include time as an independent variable. They are used for transient studies or for time-dependent physical systems.

Parabolic equations model diffusion phenomena such as temperature in a body, eddy current diffusion, or diffusion of impurities in a semiconductor device. They are a generalization of the heat equation

$$\frac{\partial u}{\partial t} - \Delta u = f_\Omega \quad \text{in } \Omega$$

initial boundary condition on Γ

$$(5.7)$$

Hyperbolic equations are a generalization of the wave propagation equation in mechanics and electromagnetism

$$\frac{\partial^2 u}{\partial t^2} - \Delta u = f_\Omega \quad \text{in } \Omega$$

initial boundary condition on Γ

$$(5.8)$$

Elliptic equations are used in steady-state problems of magnetostatic and electrostatic, solid mechanics for deformation calculations, fluid mechanics for Laplacian flows, and heat conduction for temperature distribution. They are a generalization of the Poisson equation

$$- \Delta u = f_\Omega \quad \text{in } \Omega$$

boundary condition on Γ

$$(5.9)$$

In electrostatics, u is the potential and f_Ω a function of the charges inside Ω.

Because u and f_Ω *can be irregular functions, the discretization process for the finite element method is not done directly on the differential formulation as for the Finite Difference Method, but by means of an integral formulation.* [note1: there exists also a Finite Difference Energy Method where the derivatives are directly approximated by finite differences in the integral (Bushnell and Almroth 1971)].

Two procedures are available for obtaining the integral formulation: the *Weighted residual method* and the *variational principle* (Oden and Reddy 1976).

The *weighted residual method* starts from the Boundary Value Problem $L(u) + f_\Omega = 0$ and establishes an integral

$$W(u) = \int_\Omega v^T R(u) \, d\Omega$$

$$(5.10)$$

of the residual form $R(u) = L(u) + f_\Omega$ which vanishes when u is the exact solution.

In the residual form, a vector function u of the space variable x is approximated by n linearly independent base functions $\Phi_k(x)$ or by n blending functions $N_k(x)$ (or shape functions, or trial functions), about which we will give more information in section 5.3.

$$u(x) = \Phi_k(x) \lozenge a_k$$

$$(5.11)$$

$$k = 1, ..., n$$

$$u(x) = N_k(x) \lozenge u_k$$

$$(5.12)$$

a_k and $u_k = u(x_k)$ are respectively called "algebraic and geometric coefficients" in solid modeling (Mortenson 1985) or "generalized nodeless and nodal parameters" by the finite element community (Dhatt and Touzot 1984). In solid modeling, one is only interested in geometrical entities (Stewart 1990) but in FEM one is interested in both physical and geometrical entities. Therefore we will use the terminology "nodal and nodeless parameters".

A solution of the weighted integral is dependent on the choice of the test functions (or weighting functions) v.
[note2: For generalization, in terms of distributions (see (Roddier, 1971) and books cited therein) when u is a scalar function piecewise integrable in Ω, we can define a distribution also denoted

$$\langle u,v \rangle = \int_\Omega u v \, d\Omega \text{ where } \int_\Omega \text{ is a Lebesgue integral, and v is a test function with compact support (v}$$

vanishes on $\partial\Omega$) whose infinite derivatives exist. By Green's formulae, the mth derivative of the distribution u is $<D_i^m u, v> = (-1)^{|m|} <u, D_i^m v>$].

Less restrictive than the mathematical problem in note2, a Boundary Value Problem of order m only needs sufficiently smooth functions v whose derivatives exist up to order m more.

The integration by parts using Green's formulae of the integral W(u) gives a *weak formulation* in place of the "strong" Boundary Value Problem. The weak formulation is interesting because the order of derivatives on u is diminished by one after each integration by parts while the order of derivatives on v is increased by one each time. Moreover, boundary conditions are automatically inserted in the weak formulation. [note3: Consider for example the linear elliptic problem $-\Delta u = f_\Omega$; if v is a sufficiently smooth function in Ω we can

$$\text{write} \int_\Omega u(-\Delta v) \, d\Omega = \int_\Omega v f_\Omega + \int_\Gamma \frac{\partial u}{\partial n} v \, d\Gamma - \int_\Gamma u \frac{\partial v}{\partial n} d\Gamma \text{ where the second right hand side is zero}$$

because v vanish on Γ].

Choosing n independent weighting functions $v_i(x)$, the nodal parameters or nodeless parameters approximating u are given by n equations $w_i = 0$, $i = 1, ..., n$. A different name is attached to each alternative choice of the weighting function:

Point collocation when $v_i(x) = \delta(x_i)$ where δ is the Dirac distribution. The equation is only satisfied at the n points of Ω where the residual is zero.

Subdomain collocation when $v_i(x) = 1$ in a subdomain Ω_i and zero elsewhere.

Galerkin method when $v_i(x) = \delta u$, where δ means "any variation of u". This method is interesting because it leads frequently to symmetric matrices.

The *Variational principle* studies the variations $\delta\pi$ of a functional π. In other words the variational formulation allows us to locate the set of state variables u for which a given functional is maximum, minimum, or has a saddle point. The variational principle is very useful because models of a physical system can frequently be stated directly in a variational form. For example, the total potential energy of a body in mechanics can be expressed by a functional $\pi = W + U$ where W is the potential energy of the external loads and U is the internal strain energy. Moreover, when a variational principle exists, and when $\mathbf{u}, \dfrac{\partial \mathbf{u}}{\partial x_i}$ ... in the functional π does not exceed a power of 2, the discretization process always leads to symmetric matrices.

The discretization process defining the stationarity of the functional π with respect to the approximation (5.12) is known as the Ritz Method.

$$\delta\pi \ (\mathbf{u}_1, \mathbf{u}_2,...,\mathbf{u}_n) = \frac{\partial\pi}{\partial\mathbf{u}} \ \delta\mathbf{u} = \frac{\partial\pi}{\partial\mathbf{u}_1} \ \delta\mathbf{u}_1 + ... + \frac{\partial\pi}{\partial\mathbf{u}_n} \cdot \delta\mathbf{u}_n = 0 \qquad (5.13)$$

It leads to the equations $\dfrac{\partial\pi}{\partial\mathbf{u}_i} = 0, \ i = 1, ..., n$ because (5.13) is true for any variation $\delta\mathbf{u}_i$.

A variational principle directly related to physical systems is identical to the Galerkin weighting method. In this case, the relationship $W(\mathbf{u})=0$ can be interpreted as a condition of stationarity because the solution $\mathbf{u}$ for W gives also $\delta\pi = 0$. The principle is said *natural* (Zienkiewicz 1977) because the operators in $\delta\pi$ are the operators $\mathbf{A}$ and $\mathbf{B}$ of the Boundary Value Problem,

$$d\pi = \int_\Omega \delta\mathbf{u}^T \mathbf{A}(\mathbf{u}) \ d\Omega + \int_\Gamma \delta\mathbf{u}^T \mathbf{B}(\mathbf{u}) \ d\Gamma = 0 \qquad (5.14)$$

Another kind of variational principle can be entirely *constructed* using Lagrange multipliers, penalty functions or the least squares method.

The *Lagrange Multipliers* allow us to take into account additional physical *constraints* which are new differential relationships $C(\mathbf{u})=0$ in Ω. These constraints are introduced in the original problem by adding the term

$$\int_\Omega \mathbf{m}^T \cdot C(\mathbf{u}) \ d\Omega \qquad (5.15)$$

to the natural functional π, where $\mathbf{m}$ are independent Lagrangian multipliers which may have a physical meaning.

Thus, the stationarity of the new functional π^* is

$$\delta\pi^* = \frac{\partial\pi^*}{\partial\mathbf{u}} \ \delta\mathbf{u} + \frac{\partial\pi^*}{\partial\mathbf{m}} \ \partial\mathbf{m} = 0 \qquad (5.16)$$

leading to a set of equations for the increased set of unknowns $\mathbf{u}$ and $\mathbf{m}$.

The method can be used for contact calculations between two or more bodies (Chaudhary and Bathe 1986) by adding to the functional π a new differential relationship defined on Γ in place of Ω. The method can be unstable because the new functional is not positive definite even if π is positive definite.

To ensure a positive definite functional with additional constraints $C(u)$ and without increasing the number of unknowns, the *penalty function method* introduces a functional

$$\rho \cdot \int_\Omega C^T(u) \cdot C(u) \, d\Omega \qquad (5.17)$$

whose stationarity satisfies the constraint $C(u)$. ρ is a positive penalty number which must be sufficiently large to impose satisfaction of the constraint when u is only an approximate solution.

Starting from (5.17), the *least squares method* considers the set of equations (5.1) as constraints $C(u)$ writing the functional π in the form

$$\pi = \int_\Omega A^T(u) \cdot A(u) \, d\Omega \qquad (5.18)$$

The expression can be seen as a weighted residual method where the weighting function is the residual itself. Obviously, the variation of π will be zero when the sum of the squares of the residuals is minimum.

So far the functionals were expressed with the variables u which correspond to the displacement in mechanics. A dual formulation to the displacement-based FEM is the stress-based FEM which express the functional versus the complementary stress variable. This functional is then called the *total complementary energy* , allowing a direct solution for stresses.

Other formulations called the *mixed formulation* and the *hybrid formulation*, keep both stress and displacement as variables, allowing more flexibility (Oden and Reddy 1976; Kardestuncer and Norrie 1987).

5.3. Basic principles

Domains encountered in engineering are often geometrically complex. Therefore, some difficulties arise for finding an approximate solution u by means of an integral formulation minimizing the functional π over the entire domain.

The FEM simplifies the complex problem by subdividing the domain Ω into many simple subdomains Ω_e called elements and applying one of the principles described so far to each particular element. In this way the stationarity of elementary functionals π_e is required

within each element and the total functional π is expressed as $\pi = \sum_e \pi_e$.

In analysis of structures the system is naturally discrete. In three-dimensional solids, automatic mesh generators for finite elements are given in (Ho-Le 1988) and (Yerry and Shephard 1984). Mesh generators are used with Computer-Aided Design systems employing solid geometric model representations (Razdan et al. 1989). The element size must be sufficiently small to keep the error of approximation acceptable, and the element sides must be of sufficiently similar size to avoid the well known degradation of their numerical performance.

For each element an approximated function $\mathbf{u}(\mathbf{x})$ can be constructed using the expression (12) but with all entities now defined within an element e.

$$\mathbf{u}(\mathbf{x}) = \mathbf{N}_e(\mathbf{x}) \lozenge \mathbf{u}_e \qquad e = 1,\ldots,M \tag{5.19}$$

where M is the number of nodal parameters in the element.

The approximations are constructed in such way that $\mathbf{u}(\mathbf{x})$ is a continuous function with continuous derivatives up to a given order in Ω_e, and ensuring a continuity of order m (noted C^m) between each subdomain. [note4: When the in- and inter-element continuity is of same order, the element is said *conform*. In practice satisfactory results can be observed when continuity between elements is not ensured (in bending of plates for example (Zienkiewicz, 1977)). These elements are so-called *non conforming* elements because they have a different kind of variation for the state variable along the common boundaries. Convergence is therefore ensured when no contribution to the virtual work arises at the element interface when the element size is reduced. That condition is detected by using the so-called *patch test* (Bazeley et al. 1965)].

In the remainder of this section, we will rely on examples taken from continuum mechanics, describing a three-dimensional FEM based on unknown displacements $\mathbf{u}$.

Consider first the static equilibrium of an element in a three-dimensional space. The element is submitted to external forces such as surface tractions $\mathbf{f}_s$ (pressure ...), body forces $\mathbf{f}_b$ (gravity ...), and to small imposed displacements $\delta\mathbf{u}$ satisfying the boundary conditions. The *virtual displacement principle* states that the element is in equilibrium when the external virtual work due to external forces is equal to the internal virtual work due to elastic reaction, which is a function of internal stresses τ and a function of small virtual strains $\mathbf{e}$ corresponding to the imposed small virtual displacements.

$$\int_{V_e} \delta_{\mathbf{e}}^{\mathrm{T}} . \tau \, dv_e = \int_{V_e} \delta_{\mathbf{u}}^{\mathrm{T}} . \mathbf{f}_b . dv_e + \int_{S_e} \delta\mathbf{u}^{\mathrm{T}} . \mathbf{f}_s . dS_e \tag{5.20}$$

In this expression

$$\mathbf{e} = \mathbf{L} \, \mathbf{u} \tag{5.21}$$

is the strain vector whose the six components are deduced from the Lagrangian infinitesimal strain tensor

$$e_{ij} = \frac{1}{2}\left(\frac{\partial u_i}{\partial x_j} + \frac{\partial u_j}{\partial x_i}\right) \tag{5.22}$$

$\tau = Ce$ is the stress vector in the infinitesimal displacement condition (Constitutive law, also called Hooke's law) whose the six components are deduced from the Cauchy stress tensor τ_{ij}, and C is the constant elasticity tensor.

With (5.19) and (5.21) we can introduce the strain-displacement matrix B so that

$$e = Bu_e \tag{5.23}$$

and write

$$\int_{v_e} \delta u_e^T . B^T . C . B \, dv_e . u_e = \int_{v_e} \delta u_e^T . N^T . f_b \, .dv_e + \int_{S_e} du_e^T .N^T .f_s dS_e \tag{5.24}$$

from which we deduce the *element-matrix equation*

$$K_e \, \Diamond \, u_e = r_e \tag{5.25}$$

where K_e is the stiffness matrix and r_e is the vector of acting forces.

As a link with section 2, note that the virtual displacement principle appears as the weak form of the equilibrium equations for analysis of solids when the test function is δu. i.e. we have the Galerkin method giving

$$w_e(u_e) = du_e^T \, \Diamond(K_e u_e - r_e) \tag{5.26}$$

It is also identical to the principle of stationarity of the elementary potential π_e.

Considering the equation (5.25), it is possible to calculate the matrices K_e and r_e in a *global coordinate system* X defining the domain Ω. But computations are in general carried out in another system, in order to simplify the construction of the shape functions N_e, and to simplify the integrals within each element.

Although this process is a priori time consuming because it needs *coordinate transformations*, it is always used because integrations in curved elements, expressed in global coordinates are not very efficient. There is a trade-off, very difficult to resolve, between the time spent in coordinate transformations and the precision in calculations. In iterative calculations, a poor precision can affect on the calculation time because the convergence towards the solution is slower.

In the more general formulation, one defines three coordinate axes called *global* (axes X_1, X_2, X_3 for the entire domain), *local* (also called an element coordinate system, with axes x_1, x_2, x_3 attached to each element Ω_e.), and *natural* (with normalized axes r,s,t for defining a natural element Ω_n, also called *master-element* or *parent-element or reference-element*). Depending on the dimension of the natural coordinate system, reference elements are made up of straight lines (1D), right triangles or squares (2D), right tetrahedra, prisms or cubes (3D).

The transformation T between reference and local elements involves a one-to-one mapping.

$$\text{T:} \quad x = x(r) = N_k^{'}(r) \cdot x_k \qquad k = 1, ..., N \tag{5.27}$$

where N is the number of nodes, and r stands for r,s,t, $N_k^{'}(r)$ is the geometrical shape function expressed in natural coordinates, and x_k is the coordinate of the node k.
For example, a square reference element allows cubic approximation of x when we use a complete polynomial basis $\Phi_{ij}(r,s) = r^i s^j$ such as

$$x\,(r,s) = \sum_{i=0}^{3} \sum_{j=0}^{3} a_{ij} \cdot \Phi_{ij}(r,s) \tag{5.28}$$

From this we can deduce a relation of type (5.25) with 16 nodes x_k (Figure 5.1a).

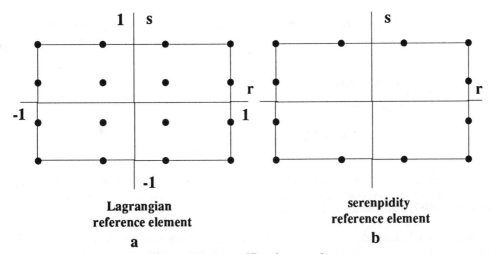

Figure 5.1. Some 2D reference elements

The reference element presented in Figure 5.1a pertains to the well known Lagrange family which gives a systematic method of generating shape functions $N_k^{'}(r)$ equal to unity at node x_k and zero at all other nodes. From this fundamental property of the shape function we note that x_k is an *interpolation function* for approximating x. Approximation functions encountered in solid modeling (Mortenson 1985) using x_k as control nodes not situated on the patch are not allowed in this context.

Reference elements created from cubic Lagrange polynomials contain a large number of terms of third degree not all of which are necessary to ensure a cubic approximation for x. Starting from an incomplete polynomial basis of 12 terms in place of 16

$$\Phi(r,s) = \begin{bmatrix} s & r & s & r^2 & rs & s^2 & r^3 & r^2s & rs^2 & s^3 & r^3s & rs^3 \end{bmatrix} \tag{5.29}$$

we suppress the 4 internal nodes without modifying the order of the approximation inside and on the element boundary. Figure 5.1b shows an element of this type which pertains to the so-called Serendipity family.

In problems of mechanics, two entities must be approximated: the displacement $\mathbf{u}(\mathbf{x})$, which is a *physical* entity and the shape of the element, which is a *geometrical* entity. Each one is given by the interpolating functions $N_e(\mathbf{r})$ and $N_k'(\mathbf{r})$. Depending on the problem, the physical and geometrical entities can be defined or not with the same nodes. Figure 5.2 shows a two-dimensional element for different orders n and m of the interpolating functions.

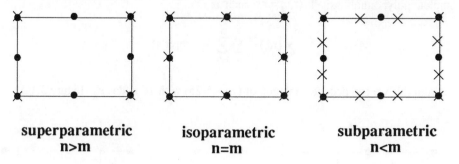

superparametric	isoparametric	subparametric
n>m	n=m	n<m

● **geometric node**

✕ **physical node**

Figure 5.2. Super- iso- and sub-parametric
two dimensional elements

An *isoparametric* element is defined by n=m and $N_e = N_k'$. It is extensively used because of its simplicity. A good summary of isoparametric finite elements is given in (Kardestuncer and Norrie 1987) and (Dhatt and Touzot 1984).

Note that all elements described so far ensure continuity of order m and n inside the element and C^0 continuity between each element. Other elements based on Hermite polynomials introduce some nodal parameters $\mathbf{u}_e$ which are derivatives. In this way, they automatically ensure continuity of order n and m inside the element and C^1 continuity between each element.

Matrices $\mathbf{K}_e$ and $\mathbf{r}_e$ in (5.25) contain derivatives of $\mathbf{u}$ and integrals over surfaces and volume expressed in the local coordinate system. They necessitate some mapping manipulations for expressing them in the natural coordinate system. From (5.19) derivatives of $\mathbf{u}$ are

$$\frac{\partial \mathbf{u}(\mathbf{x})}{\partial x_i} = \frac{\partial N_e(\mathbf{x})}{\partial x_i} \cdot \mathbf{u}_e \tag{5.30}$$

By the rule of partial differentiation we can write

$$\left[\frac{\partial Ne(x)}{\partial x_1} \quad \frac{\partial Ne(x)}{\partial x_2} \quad \frac{\partial Ne(x)}{\partial x_3} \right]^T = \mathbf{J}^{-1} \cdot \left[\frac{\partial Ne(r)}{\partial r} \quad \frac{\partial Ne(r)}{\partial s} \quad \frac{\partial Ne(r)}{\partial t} \right]^T \qquad (5.31)$$

where $\mathbf{J}$ is the *Jacobian matrix* defined by $\left[\frac{\partial x}{\partial r} \right]$ whose components are $\frac{\partial x_i}{\partial r} = \frac{\partial N_k}{\partial r}(r) \cdot x_k$.

In this way the integral

$$\mathbf{K}_e = \int_{\Omega_e} \mathbf{B}^T(x) \cdot \mathbf{C} \cdot \mathbf{B}(x) d\Omega_e \qquad (5.32)$$

is written in the form

$$\mathbf{K}_e = \int_{\Omega_n} \mathbf{B}^T(r) \cdot \mathbf{C} \cdot \mathbf{B}(r) |\mathbf{J}| \, d\Omega_n = \int_{\Omega_n} \mathbf{F}(r) \, d\Omega_n \qquad (5.33)$$

where $|\mathbf{J}|$ is the determinant of the Jacobian operator.

A numerical integration involving Newton-Cotes or Gauss quadrature can now be used. With the Gauss quadrature we can write

$$\mathbf{K}_e = \sum_{i,j,k}^{n} \alpha_{ijk} \, \mathbf{F}(r_i, s_j, t_k) \qquad (5.34)$$

evaluating the integral at the n^3 points r_i, s_j, t_k situated inside the reference element and whose position and weights α_{ijk} depend on n. (Ralston, 1965).
r_e is evaluated by a similar calculation involving surface and volume integrals with force calculations based on the same shape function $N_e(r)$ used for the stiffness coefficients (forces are called *consistent* loads in this case).

The transformation T: $\mathbf{dx} = \mathbf{J} \cdot \mathbf{dr}$ and T^{-1} : $\mathbf{dr} = \mathbf{J}^{-1} \, \mathbf{dx}$ are the inverse of one another on condition that the Jacobian does not vanish, i.e. when there exists a one-to-one mapping and continuous partial derivatives satisfying $|\mathbf{J}| > 0$ for every point in the domain. This unique relation between the coordinate systems does not exist when the element is greatly distorted.

Figure 5.3 resume coordinate transformations for cube and prism reference elements.

In addition to the transformation T, the coordinate systems involve a rotation R between local and global axes. The mapping from natural to global coordinates is often done directly avoiding the rotation step. The situation shown in Figure 3 is necessary when the order of the global coordinate system is higher than the order of the natural coordinate system. The inverse transformation T^{-1} cannot be defined because the Jacobian matrix is not a square matrix. This situation arises for trusses and plates oriented in a three-dimensional space.

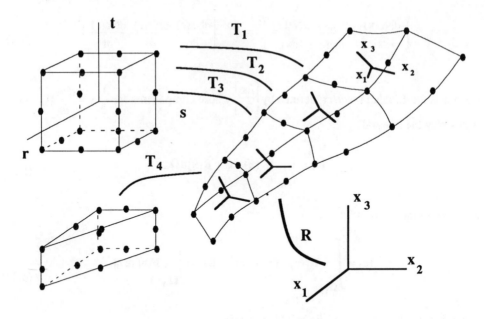

Figure 5.3. Coordinate transformations

The following step in the finite element process concerns the assembly of element-matrices, expressed in global coordinates, in a *complete element matrix* equation

$$K.U = R \tag{5.35}$$

obtained from $W = \sum_e W_e$ with W_e given in (5.26). Where K is a nxn square matrix, U and R are nx1 vectors, n the number of degrees of freedom (DOF).

Each node appearing in an element makes a contribution to the complete element matrix and must be inserted in the right place by means of an index table.

For reducing the overall size of the complete matrix K, several tricks are commonly used:

- the assembly is sometimes done through corner nodes only,
- equations giving the essential boundary conditions (prescribed displacements) are not assembled,
- a skyline storage is used ...

Using the sparsity of K, the skyline algorithm stores the matrix K column by column, each column starting from the first non zero term and ending on the diagonal. When K is a non symmetric matrix the same process is carried out with the lower triangular matrix. The matrix K is sparse because the terms coupling two nodes (more precisely two DOF) are always zero except when nodes (DOF) are common to the same element.

As the matrix sparsity is directly related to node numbering, automatic resequencing algorithms have been developed (Cuthill and McKee 1969), (Gibbs et al. 1976).

A few solution methods for the linear system (5.35) are listed in table 5.I.

5.4.   Introduction to non-linear and dynamic analysis

In this section we proceed with the most general case, extracted from mechanics, when a body is subjected to large displacements, large strains or to boundary conditions modified by contacts. Such **non-linearities** are commonly classified as *geometric* or *material*.

Geometric non-linearities are concerned with large displacements and large rotations which can arise in small or large strains. Non-linear properties are generally processed by introducing in the equilibrium relation (5.24), written at time t, the so-called 2nd Piola-Kirchoff stress tensor (components $_0^tS_{ij}$) combined with the Green-Lagrange strain tensor (components $_0^t\varepsilon_{ij}$) (Mase 1970)

$$_0^tS_{ij} = \frac{^0\rho}{^t\rho} \cdot \frac{\partial ^0x_i}{\partial ^tx_m} \cdot \tau_{mn} \cdot \frac{\partial ^0x_j}{\partial ^tx_n}$$
(5.36)

$$_0^t\varepsilon_{ij} = \frac{1}{2}\left(\frac{\partial ^tu_i}{\partial ^0x_j} + \frac{\partial ^tu_j}{\partial ^0x_i} + \frac{\partial ^tu_k}{\partial ^0x_i} \cdot \frac{\partial ^tu_k}{\partial ^0x_j} \right)$$
(5.37)

[note4: These convenient notations are defined by (Bathe 1982) with $^tx = {^0x} + {^tu}$ and $\frac{^0\rho}{^t\rho}$ the ratio of mass densities. The components are expressed at time t in relation with the configuration ^0x of the body at time 0 ($^tx = {^0x} + {^tu}$). This is the so-called Total Lagrangian formulation extensively used in analysis of solids and structures (or Updated Lagrangian formulation when components are expressed in reference to the last known configuration). The alternative Eulerian formulation describing the motion by $^0x = {^0x}\,(^tx, t)$ is generally used for analysis in fluid mechanics].

S and $\mathcal{E}$ are invariant under rigid body motion and are energetically conjugate, i.e.

$$\int_{^0V_e} {_0^tS} \lozenge \delta _0^t\varepsilon ^0dV_e = \int_{^tV_e} \frac{^0\rho}{^t\rho} \lozenge {^t\tau} \lozenge \delta _te \, ^tdV_e$$
(5.38)

Another stress-strain couple energetically conjugate and invariant is the Jaumann stress tensor associated with the velocity strain tensor (Kojic and Bathe 1987). They are appropriate for impact problems involving elastic-plastic flow (Hallquist et al. 1977).

Material non-linearities are introduced by means of the constitutive relation between stress and strain for modeling the behavior of non-linear elastic material such as rubber (Haggblad and Sundberg 1983), or non-elastic materials whose response depends on stress-strain history (Mase 1970).

In *non-linear static analysis* and in *non-linear pseudo-static analysis* (when loads are applied sufficiently slowly for considering each instant of time as a static equilibrium position), equation (5.24) is linearized and an *iterative incremental process* is initiated for each instant of time. Table 5.I lists a few solution methods.

Table 5.I: Methods for solving equilibrium equations
For more details and for other methods see
(Kardestuncer and Norrie 1987, part 4) and numerical analysis
books such as (Jennings 1977; Ralston 1965).

Static analysis		
linear equations	direct methods	Gauss elimination, Choleski's factorization, frontal, substructures
	Iterative methods	Gauss-Seidel, Jacobi
non-linear equations		Newton-Raphson modified Newton-Raphson quasi-Newton
Dynamic analysis		
linear and non-linear equations	direct methods	explicit integration central differences
		implicit integration Houbolt, Wilson, Newmark
	mode superposition (eigenproblem)	

In **dynamic analysis** the equilibrium relation (5.24) must be written at time t, and must include damping and inertia forces. In this way relations (5.25) and (5.35) become

$$M_e\ddot{u}_e + D_e\dot{u}_e + Ku_e = r_e \quad (5.39)$$

$$M\ddot{U} + D\dot{U} + KU = R \quad (5.40)$$

where D_e, D and M_e, M are respectively damping and mass matrices. $\ddot{u}_e$, $\ddot{U}$ and $\dot{u}_e$, $\dot{U}$ are nodal accelerations and velocities.

Equation (5.40) is solved using a technique of direct integration or a technique of mode superposition (Table 5.I). Direct integration techniques subdivide the time in steps Δt and express $\ddot{U}$ and $\dot{U}$ by a finite difference scheme. Mode superposition is essentially an eigenproblem expressing the system response as a series of eigenmodes.

5.5.   Conclusion

In conclusion we would like to relate this chapter to the problems of visualization and graphics simulation.

The FEM is a powerful analysis method for solving static and dynamic engineering problems, but it is also a powerful numerical method particularly suitable for generating natural models in an image simulation and animation context (Gourret et al. 1989; Pentland and Williams 1989). Figure 5.4 (see colour section) shows an example of the use of FEM for animation purposes (Gourret et al. 1989).

The use of interactive computer graphics for definition of the initial configuration (*preprocessing*) and for interpretation of calculated responses (*postprocessing*) is not the only relationship between FEM an CG. Development of *interactive adaptative* processors (Gatass and Abel 1983), acting during calculations for monitoring real-time or near real-time modal analysis of structures is also an important tool. It allows the understanding of the structural behaviour or modeling errors by direct visualization, and the correction of detected problems by direct interaction.

References

Bazeley G.P., Cheung Y.K., Irons B.M., Zienkienwicz O.C. (1965) Triangular elements in bending conforming and non conforming solutions, *Proc.Conf.Matrix Meth.Struc.Mech.*, Ohio.

Bathe K.J. (1982). *Finite element procedures in engineering analysis*, Prentice-Hall.

Bushnell D., Almroth B.O. (1971) Finite difference energy method for non linear shell analysis, *J.Comp.Struc.*, 1, pp.361.

Chaudhary A.B., Bathe K.J., (1986) A solution method for static and dynamic analysis of three-dimensional contact problems with friction, *Comp.Struct.*, 24(6), pp.855-873.

Courant R., Hilbert D. (1966) *Methods of mathematical physics*, J.Wiley

Cuthill E., McKee J.M. (1969) Reducing the bandwidth of sparse symmetric matrices, *Proc. 24th Nat.Conf.Assoc.Comp.Mach.*, ACM P69, pp.157-172.

Dhatt G., Touzot G. (1984) *Une présentation de la méthode des éléments finis*, Maloine Paris.

Gallagher R.H. (1975) *Finite element analysis, fundamentals*, Prentice-Hall.

Gatass M., Abel J.F. (1983) Interactive-adaptative, large displacement analysis with real-time computer graphics, *Comp.Struc.*, 16(1), pp.141-152.

Gibbs N.E., Poole Jr W.G., Stockmeyer P.K. (1976) An algorithm for reducing the bandwidth and profile of a sparse matrix, *SIAM J.Numer.Anal.*, 13, pp.236-250.

Gourret J.P., Magnenat-Thalmann N., Thalmann D. (1989) Simulation of object and human skin deformations in a grasping task, *Proc. SIGGRAPH'89, Comp.Graph.*, 23(4), Boston MA. pp.21-30.

Haggblad B., Sundberg J.A., (1983) Large strain solution of rubber components, *Comp.Struc.* 17(5,6), pp.835-843.

Hallquist J.O., Werne R.W., Wilkins M.L. (1977) High velocity impact calculations in three dimensions, *J.Appl.Mech.*, pp.793-794.

Ho-Le K. (1988) Finite element mesh generation methods:a review and classification, *CAD*, 20(1), pp.27-38.

Jennings A. (1977) *Matrix computation for engineers and scientists*, J.Wiley.

Kardestuncer D., Norrie H. (eds) (1987) *Finite element handbook*, McGraw-Hill.

Knowles N.C. (1984) Finite element analysis, *CAD*, 16(3), pp.134-140.

Kojic M., Bathe K.J. (1987) Studies of finite element procedures-stress solution of a closed elastic strain path with stretching and shearing using the updated Lagrangian Jaumann formulation, *Comp.Struc.*, 26(1,2), pp. 175-179.

Mase G.E. (1970) *Continuum mechanics*, Schaum, McGraw-Hill.

Mortenson M.E. (1985) *Geometric modeling*, J.Wiley.

Oden J.T., Reddy J.N. (1976) *An introduction to the mathematical theory of finite element*, J.Wiley.

Pentland A., Williams J. (1989) Good vibrations: Modal dynamics for graphics and animation, *Proc.SIGGRAPH'89, Comp.Graph.*, 23(4), Bostan MA, pp.215-222.

Ralston A. (1965) *A first course in numerical analysis*, McGraw-Hill.

Ratnajeevan S., Hoole H. (1989) *Computer Aided analysis and Design of electromagnetic devices*, Elsevier.

Razdan A, Henderson M.R., Chavez P.F., Erickson P.A. (1989) 'Feature based object decomposition for finite element meshing', *Visual Computer*, 5, pp.291-303.

Roddier F. (1971) *Distributions et transformation de Fourier*, Ediscience Paris.

Stewart N. (1990) Solid modelling. In: D.Thalmann (ed) *Scientific Visualization and Graphics Simulation*, Chapter 4, John Wiley and Sons (This volume)

Yerry M.A., Shephard M.S. (1984) Automatic three-dimensional mesh generation by the modified octree technique, *Int.J.Num.Meth.Eng.*, 20,pp.1965-1990.

Zienkiewicz O.C. (1977) *The finite element method*, Third edition. McGraw-Hill.

6

Visualizing Sampled Volume Data

Jane Wilhelms
University of California, Santa Cruz

6.1. Introduction

Simulations and scientific experiments often produce data in the form of a large number of sample values within a coordinate space. In numeric form, this information is often hard to comprehend, but by visualizing it, both the imaging methods of 3D graphics and the advanced information processing abilities of the human visual system can be brought to bear. For visualization, the data must eventually be mapped into color or intensity values for pixels making up a two-dimensional image. The approaches that have been taken to visualization of sampled volume data are the subject of this chapter.

Section 2 will consider the forms which sampled volume data may assume. Section 3 discusses ways in which the sample values can be mapped to visual parameters such as color, intensity, surface properties, and semi-transparency. Sections 4 and 5 discuss two prominent approaches to visualizing volumes: viewing two-dimensional slices within the volume and extracting geometrical entities from it. Section 6, the main thrust of the chapter, considers the methods that have been used to directly visualize volume data in its three-dimensional form without intermediary geometric representations. Section 7 examines some issues of importance for future research.

6.2. Sampled Volume Data

Volumetric data consists of one or more scalar and/or vector values found in an n-dimensional space. In the simplest case, the data consists of single scalar values located in a three-dimensional space. However, this space easily becomes four-dimensional if the data changes over time, and may be of much higher dimensionality for more abstract problems (Udupa et al. 1982). Often multiple parameters occur at the sample position within this space, some of which are scalar and some vector. For example, data from computational fluid dynamics may consists of a scalar for temperature and vectors for velocity. Imaging methods that can show multiple parameters in a comprehensible way are needed (Hibbard and Santek 1989).

The sample points may be distributed in various ways as well (Sabin 1986, Gelberg et al. 1989). The simplest case is a regular cubical or rectangular grid with equal or variable spacing along each axis. Such a format is common to medical imaging data. Simulations may produce samples in a warped grid format and experimental data may take the form of random sample points. Non-rectangular but regular sampling (e.g., where the samples decompose space into tetrahedrons) can actually be advantageous (Upson 1986), helping to avoid ambiguities in the interpolation of data (Bloomenthal 1988). The algorithms discussed in this chapter generally assume rectangular-gridded data, unless otherwise noted.

6.2.1 Manipulating Sampled Data

It may be convenient or necessary to manipulate the volume data before imaging. Most visualization algorithms are designed for rectangular grids, and the data may be resampled (from the samples) to this format (Frieder et al. 1985, Drebin et al. 1988, Levoy 1988). Producing a continuous image from the samples itself requires interpolation. However, for resampling to be done accurately, it must be possible to reconstruct the original continuous function from which the samples were taken, i.e. the sample frequency must be above the Nyquist frequencz (Pratt 1978). Unfortunately, it may not be possible to resample from the original continuous source, and there may be no reason to assume the sampling rate is sufficient. When interpolating and resampling are done, simple trilinear interpolation is often used (Höhne and Bernstein 1986; Cline et al. 1988; Levoy 1988; Upson and Keeler 1988). This may produce undesirable results (Van Gelder and Wilhelms 1989), though it has been suggested that it introduces less subjective interpretation into the data (Upson and Keeler 1988). Higher-order interpolation is also possible, at greater expense.

Sampled data may contain noise and artifacts, and image processing methods such as contrast enhancement and filtering may be applied (Drebin et al. 1988; Levoy 1988). Low-pass filtering the samples (removing the high-frequencies representing rapid changes in data values) (Drebin et al. 1988; Westover 1989) can remove noise and produce more visually attractive images. It might, however, alter the data in undesirable ways, e.g. interpreting accurate but extreme data values as noise.

6.2.2. Subvolumes

The subdivision of the volume into sampled regions can be viewed in two ways. First, the data associated with a particular location in the space may represent a point sample or an average of the values over a particular region in space. The region of space that this value represents is typically called a *voxel*, though the term is most commonly used in the case of rectangular grids. The data value associated with a voxel will be assumed to be constant throughout the region the voxel represents.

Alternatively, the sample points can be considered as points. For rectangular grids, eight neighboring sample points define a *computational cell* (Upson 1986; Upson and Keeler 1988). In this case, the volume is considered to consist of cell regions whose internal values vary continuously depending upon the values of the eight vertices defining the cell (Lorensen and Cline 1987, Cline et al. 1988, Levoy 1988, Upson and Keeler 1988). Cells are used when values will be interpolated between samples, though some algorithms that use voxels interpret the voxel regions as inhomogeneous by other methods (Drebin et al. 1988). For purposes of succinctness, this chapter will use the term *subvolume* as a general term for both voxels and cells.

6.2.3. Data Size

The sheer amount of sample data is often a problem. A 64x64x64 scalar voxel array will take a quarter million memory locations. If each location requires just four bytes, this is a megabyte of data. Much larger files are common. There are two problems associated with data size: first, storing the data on a reasonably accessible medium; and, second, accessing the data in a reasonable amount of time. Computers with significant disk storage are needed to store the volume data. Even if the data is available on disk, the main memory of the computer is often not sufficiently large to retain it during processing, and, thus, it must be paged on and off disk. Paging time can easily come to dominate the algorithm used, so that the system will thrash.

Three approaches might help avoid these problems: *1. Store and access the data* so that a sufficiently small amount is kept in memory at one time and paging is kept to a minimum;. *2. Compact the data,* either by simple data compression techniques or more sophisticated approaches; *3. Use hierarchical data structures* to help minimize the need to store all the data (Srihari 1981; Meagher 1982; Mao et al. 1987).

6.3 Visual Representations

While some data samples may inherently represent visual phenomena, such as CT-scans, often the data are not visual, such as temperature, pressure, or velocity data. An important point distinguishing scientific visualization from computer graphics rendering natural scenes. As the aim of scientific visualization is to communicate information about the data (not create attractive images, though that often also happens),. the relationship between the values at volume sample points and the intensity or color at the pixel is entirely flexible and need not bear any resemblance to true visual phenomena.

An important issue in scientific visualization is how to map sample values to a digital image form. Often simple *transfer functions* in the form of a table are used, for example, mapping temperature to a range of colors between blue and red. More complex mappings are also being explored (Levoy 1988; Drebin 1988; Sabella 1988). Mappings should be appropriate for the nature of the data. Choice of the colors used to visualize data is an issue of interest (Meyer and Greenberg 1980)

6.3.1. Color, Opacity, and Shading

To avoid repetition, this chapter will use *intensity* to refer to a single value for greyscale or three values for color. While most modern graphics machines are capable of color, some application (such as biomedical images) may prefer the use of grey to avoid introducing interpretations and artifacts in the image (Smith 1987).

The intensity at each pixel of the image is found by considering the contributions of all the regions of space that project onto the pixel. It is sometimes useful to treat the subvolumes, or geometrical objects defined in the volume, as semi-transparent, so that those in front do not completely occlude those farther away. This is usually done by associating an *opacity* (or *alpha* value between 0 (transparent) and 1 (opaque) with the subvolumes or objects, as well as the intensity value(s). Compositing these intensities using opacity values is discussed in Section 6.6.4 (Porter and Duff 1984).

The total intensity contribution of a subvolume or object is determined by three contributions: the light it emits; the light it transmits; and the light it reflects. The shading models traditional to computer graphics model light as coming from three sources: *ambient light* dependent only on the color of the background light and the surface reflectivity; *diffuse light* from a particular source which reflects off in all directions proportionate to the angle of incidence of the light with the surface; and *specular light* from a particular source which reflects off mostly in the direction of reflection. Shading models are described in any graphics text (Foley and Van Dam 1982).

Geometric representations, such as polygons, have well-defined surfaces with normal vectors indicating their direction. When working with volume data, surfaces are sometimes detected by using the *gradient* or variation in neighboring sample values as an indication of the surface normal. Generally, gradients are approximated by using the finite difference between neighbors on either side (Lorensen and Cline 1987, Levoy 1988) of a sample point:

$$\nabla f(x,y,z) = \quad [\ 1/2\ [\ f(x{+}1,\ y,\ z) - f(x{-}1,\ y,\ z)\ ], \qquad (6.1)$$
$$1/2\ [\ f(x,\ y{+}1,\ z) - f(x,\ y{-}1,\ z)\ ],$$
$$1/2\ [\ f(x,\ y,\ z{+}1) - f(x,\ y,\ z{+}1)\ ]\ ]$$

6.3.2 Visualizing More Complex Data Parameters

Often volumes include multiple scalar and vector values at each voxel. A combination of techniques can be used, but there is a danger of the image becoming confusing. An extra parameter may also be visualized by using texture mapping (Peachey 1985, Robertson 1985, Upson and Keeler 1988). Multiple values can be mapped to one display parameter, for example, the color may represent a combination of temperature and concentration. Icons may also be used (Grinstein et al. 1989, Stettner and Greenberg 1989).

Vector data can be represented by the use of added features and icons, such as a *hedgehog* of lines from the voxels, representing vector direction and magnitude, or *streamlines* (ribbons), which have the advantage that they can show curling well. If animation is possible, particles moving along the vector lines (*particle advection*) can be used. These methods have been discussed in (Upson 1986, Shirley and Neeman 1989, Fowler and Ware 1989, Gelberg et al. 1989, Helman and Hesselink 1989, Stettner and Greenberg 1989).

It is possible to augment the limited number of visual parameters that can be effectively shown in one image with other forms of sensory input. The frequency or rate of repetition of a sound could be used to indicate density as the user interactively moves a cursor through the voxel space (Grinstein et al. 1989). A force-feedback joystick could also represent density by resisting motion according to its magnitude.

6.4. Slicing and Carving

Slicing and carving are the simple methods of visualizing three-dimensional volumetric data, which require little computation and analysis but produce less sophisticated images.

In the simplest case, one can visualize a two-dimensional planar slice through volume (see Figure 6.1 and 6.2, color section). If the slice is orthogonal to the volume, a planar slice is produced by fixing one coordinate (e.g., *X*) and stepping through all possible values of *Y* and Z for that particular X. Visualization can be trivial, requiring only that data values in the slice be mapped to intensity, or shading models can be used (Sabella 1988). The image may show constant value voxels (giving a blocky appearance) or interpolation across cells (for continuous intensity variation). Planes may also be of an arbitrary orientation, at slightly great cost.

In some cases, it may be desirable to cut away parts of the volume with more complex operations. This can be done by applying boolean set operations (intersection, union, and difference) to the original volume plus a matte volume determining subvolumes of interest (Farrell et al. 1984, Drebin et al. 1988). Alternatively, carving can be done by calculating the the intersection with a suitably shaped carving model, such as a cylinder (Frieder et al. 1985), and removing parts outside the model. Carving may be followed by visualization using geometric representations or direct volume rendering, as described next.

6.5 Geometric and Topological Representations

A second approach to visualizing a volume of data is to extract from it geometric, and sometimes topological, features. Geometric representations describe the locations of features using points, lines and curves, surfaces, and solids. Sometimes geometric representations are accompanied by topological representations describing features as components, connections, and holes (Artzy et al. 1981; Srihari 1981; Udupa et al. 1982) which make it easier for further analysis can be done, such as finding surface area, and volume (Cook et al. 1983; Herman and Udupa 1983). Use of these representations for feature extraction from volumes has been subject of many technical articles, some of which are referenced here.

Geometric and topological features are distinguished using *threshold* values that distinguish one region of the volume from another. Algorithms using thresholding may examine each subvolume in sequential order (Lorensen and Cline 1987, Cline et al. 1988), or they may use boundary tracking to follow a boundary through the data (Liu 1977; Artzy et al. 1980; Cline et al. 1988; Pizer et al. 1986).

There are a number of advantages to using geometric representations. Modern graphics hardware is designed to manipulate and display geometric representations quickly, often allowing interactive manipulation. Most graphics techniques for visual realism are designed for geometric representations. Geometric representations *may* be more compact than the original volume. Complex volumes can be made more comprehensible by limiting the visualization to a few surfaces.

However, there are problems as well. Most of the volume data may be ignored. Thresholding requires a binary decision as to what is inside and outside a feature, and errors in classification can produce spurious positives or holes (Dürst 1988, Levoy 1988, Van Gelder and Wilhelms 1989). It may be difficult to find threshold values to distinguish the desired boundaries.

6.5.1. Cuberilles and Isosurfaces

Cuberilles and *isosurfaces* are the two main approaches used to represent sampled volumetric data as geometric entities.

Cuberilles are a simple but effective method of visualizing regions of interest (Herman and Liu 1979; Artzy et al. 1981; Herman and Udupa 1983; Chen et al. 1985). They are usually used with regular grids, where the voxels are seen as representing a region of space with a constant value, such as in CT and NMR images. Cuberilles create a binary volume consisting of voxels belonging to the object of interest, and, thus, can be used to represent only the surface, or the interior as well. They can be succinctly stored by an octree representation (Meagher 1982; Meagher 1984). Imaging is usually done by rendering the outer faces of the cuberilles as polygons. The blockiness of the image can be reduced (if desired) by shading methods (Chen et al. 1985; Hoehne and Bernstein 1986; Trousset and Schmitt 1987).

Isosurfaces represent the location of a surface of a specific threshold value within the volume. Initially, isosurfaces were found by tracing two-dimensional outlines of the contour surface on adjacent planes of rectangular grid data and joining the contours to produce a three-dimensional surface (Keppel 1975, Fuchs et al. 1977, Christiansen and Sederberg 1978, Sunguroff and Greenberg 1978; Wright and Humbrecht 1979, Ganapathy and Dennehy 1982, Cook et al. 1983, Vannier et al. 1983, Pizer et al. 1986). Other algorithms do thresholding in three-dimensions from the start (Liu 1977; Artzy et al. 1980). Marching cubes or dividing cubes are recent popular techniques which find triangles or points representing the isosurface (Cline et al. 1988, Lorensen and Cline 1987). Figures 6.3 and 6.4 (see color section) show images consisting of triangles generated by a marching cubes type of algorithm. Similar methods of finding surfaces of a given value have been used in modeling with implicit surfaces (Norton 1982, Wyvill et al. 1986, Bloomenthal 1988). Isosurfaces are usually found by interpolating between sample values, which can lead to topologically ambiguous surfaces (Van Gelder and Wilhelms 1989).

6.6. Direct Volume Rendering

Direct volume rendering refers to methods that create an image from the three-dimensional volume without generating an intermediate geometrical representation. An intensity/opacity contribution for each subvolume onto each pixel is calculated directly from the volume data. This approach has excited considerable interest in the computer graphics community (Smith 1987, Drebin et al. 1988, Levoy 1988, Sabella 1988, Upson and Keeler 1988, Westover 1989). Figures 6.5 and 6.6 (see color section) were created using a ray-tracing direct volume renderer.

Direct volume rendering is a natural extension of earlier voxel and cuberille methods, the chief differences being the direct method's emphasis on the underlying continuous nature of the sampled volume, and its avoidance of geometric models in the rendering process. Early research into direct volume rendering include (Harris et al. 1978; Farrell 1983; Gibson 1983; Vannier et al. 1983; Farrell et al. 1984; Tuy and Tuy 1984; Frieder et al. 1985; Schlusselberg et al. 1986; Hoehne and Bernstein 1987; Trousset and Schmitt 1987). Direct volume rendering has also been used for realistic imaging to simulate the scattering of light (Blinn 1982; Kajiya and Von Herzen 1984; Max 1986).

Recent work is notable for the extreme flexibility of the mappings from data to visual parameters, and the use of semi-transparency for continuous images together with thresholds for distinguishing features (Smith 1987; Drebin et al. 1988; Levoy 1988; Sabella 1988; Upson and Keeler 1988; Westover 1989).

The two main approaches used for volume rendering are *ray tracing* (Tuy and Tuy 1984; Schlusselberg et al. 1986; Levoy 1988; Sabella 1988; Upson and Keeler 1988) and *projection* (Upson and Keeler 1988; Westover 1989). Both of these are methods in common use for rendering geometric data as well (Foley and Van Dam 1982).

.   In *ray tracing*, rays are cast out from the viewer through the hypothetical screen pixels into the three-dimensional volume. The contribution to the pixel of the voxels or cells that the ray meets are calculated and used to find a color or intensity for the pixel. Fundamentally, the algorithmic loop is:

```
for each ray do
     for each voxel-ray intersection do
          calculate pixel contribution
```

.   In *projection methods*, each voxel or cell is projected onto the hypothetical screen and its contribution to the pixels it affects is calculated. In this case, the loop is

```
for each voxel or cell do
     for each pixel projected onto do
          calculate pixel contribution
```

Direct volume rendering is attractive for a number of reasons. It can image the volume as a continuous material and also isolate important data features within it (Upson and Keeler 1988). Subtle variations throughout the volume, which cannot be represented well with surfaces alone, can be visualized using semi-transparency (Schlusselberg et al. 1986; Drebin et al. 1988). Sophisticated mapping can bring out complex data characteristics (Drebin et al. 1988, Sabella 1988). Volume representations can be combined with geometric data with relative ease (Levoy 1989b). Binary classifications about what is and isn't part of a feature need not be made. Often volume data models are inherently fuzzy and boundaries are transition zones (Smith 1987; Levoy 1988), which the fuzzy quality of semi-transparent images effectively visualizes. While the size of volume data may be enormous, the amount of memory taken up by isosurfaces generated from the volumes may be even greater (Drebin et al. 1988; Upson and Keeler 1988).

Volume rendered images have their drawbacks, however. Semi-transparent images may be difficult to interpret. Slight changes in opacity mappings or interpolation may undesirably alter the image (Levoy 1988). Volume rendering is also very slow. Generally, all subvolumes must be examined for each image, and images from different viewing angles require massive recalculation. Approaches that provide interactive speeds are needed, because the ability to view the volume from arbitrary positions is necessary to interpret the image, and because acceptable mapping parameters are difficult to find and usually involve user interaction (Westover 1989).

A number of techniques can be used to speed the rendering. Mapping to visual parameters is often done using a table look-up operation. At a cost in space, mapping information, such as gradients, which can be used for multiple views can be precalculated and stored (Levoy 1988; Drebin et al. 1988; Upson and Keeler 1988; Westover 1989). At a cost in resolution, coarser renderings can be done and progressively refined (Bergman et al. 1986; Westover 1989). The regularity and simplicity of the approach lends itself well to parallel processing (Upson and Keeler 1988; Levoy 1989a; Westover 1989).

Several different but related approaches to volume rendering have recently been published (Drebin et al. 1988; Levoy 1988; Sabella 1988; Upson and Keeler 1988; Westover 1989). In an attempt to systematically examine these approaches, volume rendering will be considered as consisting of four steps: 1) geometric and viewing transformations; 2) mapping from data values to visual parameters and shading; 3) finding subvolume contributions using ray-tracing or projection approaches; 4) compositing subvolume contributions to find pixel values. These steps do not necessarily have to occur in this order, and the order in which they occur is related to the trade-offs of space versus time so common to computer algorithms.

6.6.1 Geometric and Viewing Transformations

Mapping positions in the volume to positions in screen space is simple when the volume is a regular grid (Westover 1989). The rotation matrix relating the two frames indicates how incremental changes in one frame cause incremental changes in the other. Only one volume position need be transformed using a matrix multiplication. Later positions can be incrementally determined using a *digital differential analyser* (Foley and Van Dam 1982; Fujimoto et al. 1986), the increments being provided by the rotation matrix.

There are two ways of dealing with the difference between the viewing space and the volume space. The most common is to use the known relationship between the two to map points or rays in one to the other (Levoy 1988; Snyder and Barr 1987; Upson and Keeley 1988). Another is to transform the volume and resample the data to produce a simpler mapping between the two (Drebin et al. 1988). For example, in an arbitrary viewing situation, a subvolume may map to more than one pixel and its projection may not be rectangular or, if a ray is sent through space it may pass between sample points. If the data is appropriately resampled, subvolumes can be made to map to a single pixel, and rays always to intersect sample points (Drebin et al. 1988).

Many volume renderers use parallel projection (Levoy 1988; Westover 1989), though some also offer perspective viewing (Drebin et al. 1988). Parallel projection is less computationally expensive and may be preferable for scientific data because it does not alter size with distance. Perspective projection can help clarify depth ordering.

6.6.2. Mapping to Visual Parameters

The sample values must be mapped to visual parameters for imaging. Mapping may be as simple as relating density values to a greyscale, or may involve complex intermediate parameters, texture maps, and sophisticated shading. Independent transfer functions for the different visual parameters, such as light emission, reflection, and opacity, can be used to present the maximum amount of information in one volume (Levoy 1988; Upson and Keeler 1988; Westover 1989). Mapping is generally done using table lookup for speed (Levoy 1988; Upson and Keeley 1988; Drebin et al. 1988; Westover 1989). In most cases, mapping finally results in an intensity and opacity value for each subvolume, which are accumulated and composited as described in Sections 6.6.3 and 6.6.4.

6.6.2.1 Shading

Subvolumes may be light transmitters, emitters, and/or reflectors. Light transmission with attenuation is necessary to create the semi-transparent quality important to volume rendering. The shading model may only consider environmental light sources and ambient light (Upson and Keeler 1988; Levoy 1988), or may also include light emitted by the volume itself (Drebin et al. 1988; Sabella 1988; Westover 1989).

Light reflection is treated in the manner common in computer graphics (Phong 1975; Foley and Van Dam 1982), as the sum of diffuse (Drebin et al. 1988; Levoy 1988; Upson and Keeler 1988; Westover 1989), specular (Drebin et al. 1988; Levoy 1988; Westover 1989), and ambient light (which may take the place of emitted light (Levoy 1988; Upson and Keeler 1988)). Shading parameters that do not vary can be precalculated and stored with the subvolumes, at the cost of additional memory usage. Diffuse and specular contributions require a surface normal, which is usually provided by the gradient direction. There may be also an atmospheric attenuation due to distance (Drebin et al. 1988; Sabella 1988; Upson and Keeler 1988).

6.6.2.2. Feature Extraction

Volume rendering need not result in continuous, fuzzy, semitransparent images; it can be used to bring out surfaces and other features within the volume as well (see Figure 6.6, color section). Simple feature extraction can be done by using the transfer function table to bring out values of interest (Upson and Keeler 1988). Diffuse and, particularly, specular lighting can also be effective in bringing out surfaces. The magnitude of the gradient vector is sometimes used as an indication of the presence of a surface (Drebin et al. 1988; Levoy 1988; Westover 1989). Westover finds that modulating opacities with a ramp which is indexed by gradient magnitude effectively brings out surfaces in the image. Levoy (1988) found the use of a window tended to produce thick surfaces that interfered with each other and a small window was liable to produce holes. Drebin et al (1988) use the gradient of a calculated density parameter as a surface indicator; the magnitude of the gradient is used to scale the reflected color.

6.6.2.3. Material Percentages

Some researchers, working with biomedical data, have found it useful to distinguish different materials within the subvolume (Levoy 1988; Drebin et al. 1988). Material classification may be more successful in determining region boundaries than simple thresholding because it may detect the presence of the material of interest within a subvolume whose actual value is below the threshold (Levoy 1988).

It is necessary to make certain simplifying assumptions for the material percentages method to work. Levoy (1988) assumes that a subvolume contains at most two types of materials and uses a simple linear interpolation to estimate percentages in subregions with values intermediate between values expected for the constituent materials. For this to be successful, a further assumption must be made that two materials whose range of values are contiguous in a histogram are the only ones that will share a voxel. If this is not the case, for example, if a voxel had value a which is associated with a particular material, the value a may also be due to the voxel being a composite of two materials with values b and c whose average is a.

Drebin et al (1988) use a maximum-likelihood probabilistic classifier to determine the likelihood that a particular sample value represents a particular mixture of materials. While this approach does not limit the number of materials per subvolume to two, it still requires the assumption that materials present in a subvolume must be adjacent in the histogram.

6.6.2.4. Mapping Field Contributions

Sabella (1988) in his ray-tracing algorithm uses a more unusual intermediate mapping before going to color and opacity. The volume is modeled as a field of light-emitting particles (small enough for density to be modeled as a continuous function), a simplification of light scattering models used to model natural phenomena such as clouds (Blinn 1982; Kajiya and Von Herzen 1984; Max 1986) and particle systems (see Sect. 7.2) (Reeves 1983; Reeves and Blau 1985). A ray is sent through the volume of particles accumulating the contribution of light from particles it passes while subtracting an appropriate amount due to the particles occluding light coming from behind. When light sources are present, the brightness is further attenuated by the dot product of the density gradient and the light source direction (diffuse reflection).

Four parameters are accumulated for each ray: accumulated attenuated intensity (similar to other algorithms), maximum density value along the ray, distance of maximum along the ray, and centroid of density along the ray. This choice of parameters is useful for identifying extrema, such as hot spots in the data. Mapping to color is flexible. Typically, the value component is a function of the light intensity, and the hue component associated with the peak along the ray. Saturation provides a depth cue, by creating the appearance of a fog in front of other regions of interest.

6.6.3. Determining Subvolume Contributions

Given a particular mapping algorithm, it is necessary to find the contribution of each subvolume to the final image. The two main approaches to this method, as mentioned before, are ray tracing and projection, but within these two categories there has been considerable variation in approach. A distinguishing feature of importance in finding subvolume contributions is whether a region is treated as a constant value voxel (Drebin et al. 1988; Sabella 1988; Westover 1989), or a varying cell (Levoy 1988; Upson and Keeler 1988). Treating a subvolume as a cell requires interpolation and may be more costly. On the other hand, if the volume has a relatively coarse resolution relative to the resolution of the screen it covers, the lack of interpolation may result in a blocky image.

There are advantages and disadvantages to both ray tracing and projection. Full ray tracing is much respected in computer graphics for its ability to render spectacular lighting effects, such as refraction and inter-object reflections (Whitted 1980). However, for rendering volumes, a simpler form, *ray casting* is usually used, in which recursive rays are not sent. For scientific visualization, it may be more the ease of implementation and the speed of the algorithm that makes ray tracing attractive (Levoy 1988; Upson and Keeler 1988; Levoy 1989a). Ray tracing is a point sampling method, and is due to have aliasing problems which may have to be handled by supersampling with multiple rays per pixel and/or stochastic sampling (Cook et al. 1984; Lee et al. 1985). Ray tracing has also been considered less amenable to parallel processing because there is no simple way to determine which voxels a particular ray can intersect. Thus, if individual rays are assigned to different processors, each of them may need access to the full volume of data (Drebin et al. 1988; Upson and Keeler 1988; Westover 1989).

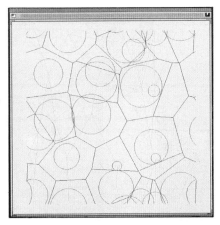

Figure 2.1. Laguerre tessellation corresponding to a family of circles on the unit torus, and representing a plane cut of a polycrystal with periodic boundaries.

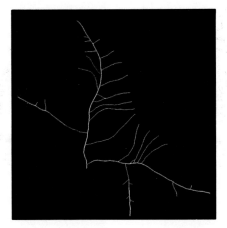

Figure 2.2. Output of a mycelia growth simulation. Successive generations of hyphae are shown in different colours.

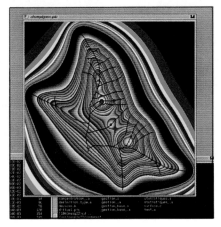

Figure 2.3. Output of a mycelia growth simulation with concentration contour lines of inhibitory substance at a given point in time.

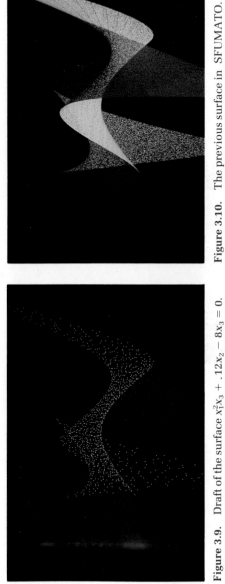

Figure 3.10. The previous surface in SFUMATO.

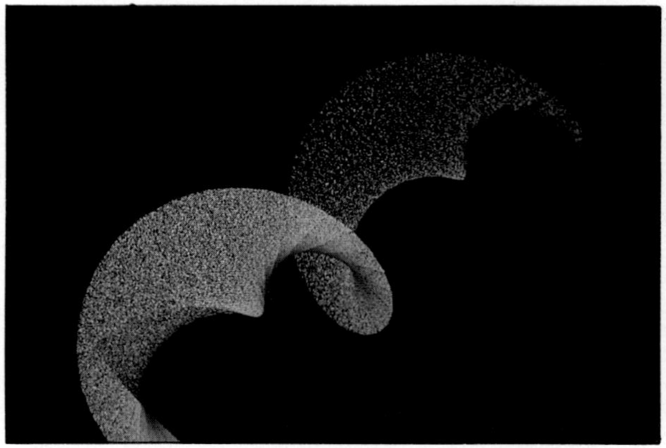

Figure 3.9. Draft of the surface $x_1^2 x_3 + .12x_2 - 8x_3 = 0.$

Figure 3.12. A space curve.

Figure 3.11. A surface of constant mean curvature in SFUMATO.

Figure 3.13. Stripes.

Figure 3.14. Stripes.

Figure 5.4. Grasping of a ball submitted to internal pressure (from Gourret et al. 1989).

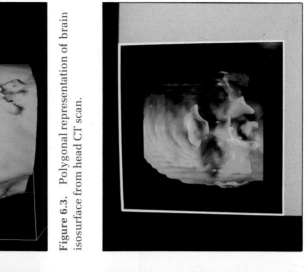

Figure 6.3. Polygonal representation of brain isosurface from head CT scan.

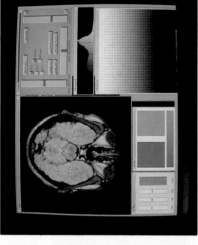

Figure 6.2. Slice through CT scan of head —interpolating cell values.

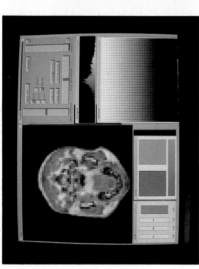

Figure 6.1. Slice through CT scan of head —shading voxels.

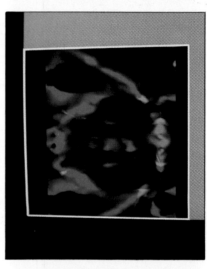

Figure 6.6. Direct volume rendering of pelvic CT scan extracting isosurface.

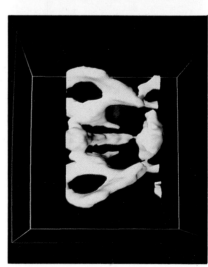

Figure 6.5. Direct volume rendering of head CT scan using semi-transparency.

Figure 6.4. Polygonal representation of bone isosurface from pelvic CT scan.

Figure 7.1. Natural maple leaf and synthesized fractal approximations using a mean square error criterion.

Figure 7.2. Composite image (created by J. Lévy Véhel).

Figure 7.7. "Natural" palm tree reconstructed from a botanical model (from Eyrolles 1986).

Figure 7.14. Evolution of a maple leaf using the free surface evolution technique (from Lienhardt 1987).

Figure 7.8. Weeping willow obtained from the same program (from Eyrolles 1986).

Figure 7.15. Bell flower obtained by Lienhardt's technique (from Lienhardt 1987).

Figure 8.4. Parametric keyframe animation: key positions are in yellow, inbetweens are in red.

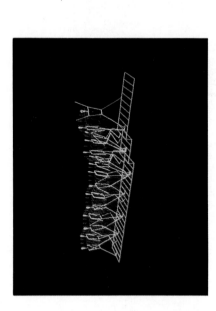

Figure 8.9. Surface mapping using JLD operators.

Figure 8.11. Phoneme expressions from the film "Rendez-vous à Montréal".

Figure 8.10. Surface mapping and hand deformations from the film "Rendez-vous à Montréal".

Figure 9.1. Object grasping sequence.

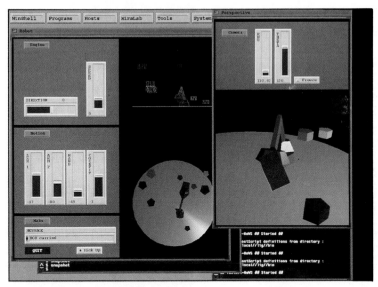

Figure 9.7. Manipulation of a robotic arm using the FIFTH DIMENSION Toolkit (Turner et al. 1990).

Figure 10.2. Example of an unstructured mesh (triangular facets extracted from tetrahedral mesh).

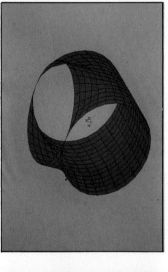

Figure 10.3. Example of a structured mesh (hexahedral cells). The model is decomposed in 2 subdomains.

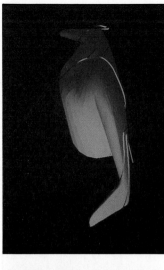

Figure 10.6. Continuously shaded scalar values (densities) on parts of the finite element model and on a plane intersection.

Figure 10.7. Example of scalar contour line plot (pressures) on parts of the finite element model and on a plane intersection.

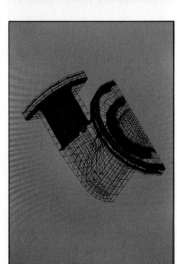

Figure 10.8. Example of iso-surface plot. The solidification front in a mould is visualized by 3 iso-surfaces corresponding to constant temperature values. Computation by Droux (1988).

Figure 10.9. Example of particle tracks in a flow field. Pressure values along the track are represented by continuous color tones.

6.6.3.1. Ray Tracing

The ray tracing algorithm involves first defining a ray penetrating the volume, then stepping along it to find contributions of subvolumes, and repeating with the next ray until the whole volume is considered. Subvolume contributions depend upon the mappings used to find visual parameters (described in the previous section) as well as how their contributions are evaluated (described here).

In order to examine ray/volume intersections, both must be defined in the same coordinate space. This is usually accomplished by transforming the rays to volume space (Levoy 1988; Sabella 1988; Upson and Keeler 1988). Rays can be sent through each pixel center, or through pixel corners which are averaged to find the pixel value, or multiple rays can be sent through each pixel.

Once the ray is defined, it is necessary to step along it accumulating contributions from the volume. This can be done by taking equidistant steps along the ray, or by calculating intersections with subvolume boundaries. Levoy (1988) took equidistant steps through subvolumes treated as interpolated cells. For each intersection point, color and opacity at the eight corners of the cell where it occurs are found and trilinear interpolation used to find the values at the ray intersection point.

Alternatively, one can step along finding intersections with subvolume boundaries and finding the contribution of the regions between the intersections (Upson and Keeler 1988). Intersections can be found relatively simply because of the regularity of the volume grid, using a digital differential analyser (Fujimoto 1986). Once one intersection is found, intersections with all further subvolumes can be calculated by using the slope of the ray.

If the volume is treated as homogeneous voxels, the distance between the entry and exit intersections of the ray with the voxel can be used to find that region's contribution (Kajiya 1984; Sabella 1988). If the volume is treated as cells, interpolation can used. Upson and Keeler (1988) calculate a trilinear function representing the variation of the shading, opacity, and texture map values through the cell and then use variable-stepsize trapezoid-rule quadrature (Press et al. 1986) to estimate the light and opacity along the ray.

New rays can be calculated by differencing, using the rotation matrix defining the relationship between the volume and viewing coordinate frames.

6.6.3.2. Cell Projection

For cell projection, the subvolumes are considered one by one, deciding which pixels of the screen they project to and accumulating their effect (Drebin et al. 1988; Upson and Keeler 1988; Westover 1989). Many researchers (Drebin et al. 1988; Upson and Keeler 1988; Westover 1989) have preferred this technique because it can provide superior antialiasing, is more amenable to parallel processing, and can use coherence in the rendering process. Three somewhat different approaches to projection have been used (Drebin et al. 1988; Upson and Keeler 1988; Westover 1989).

Drebin et al. (1988) interpret subvolumes as voxels, but though they don't use interpolation, they achieve somewhat the same effect through their use of material characteristics (Drebin et al. 1988). If just one material, or more than one material with the same density, is found in the voxel, no surface is recognized and the color of the voxel is that emitted by its material. However, if a density variation exists, the density gradient is used to determine which material is in front, and intra-voxel compositing is then used (see Section 6.6.4), to mix the front light emitting material, the reflective surface, followed by the back light emitting material. Actual compositing is used to avoid color bleeding and make surfaces more distinct. Drebin et al. (1988) resample the volume dependent upon the eyepoint so that a particular voxel projects only to one pixel, making this operation trivial.

Upson and Keeler (1988) interpret subvolumes as cells and use trilinear interpolation to visualize regions between sample points, as in their ray-tracing approach. They use their knowledge of the relative positions of the viewer and the volume to determine the order to step through cells front to back. For each cell, a bounding box is determined and clipped to scanlines to produce pixel runs that the cell may affect. For each scanline, a polygon defining the cell's extent is found. This polygon is broken down into (at most five) spans which are regions where the same two faces form the front and back boundary of the portion of the projected cell. Spans can be processed using vectorization. Trapezoid quadrature is used at the pixel corners to find light and opacity contributions of the cell there, and these are averaged across the pixel for a final value. This method is a volumetric analog of the Z-buffer, and, thus, they call it the *V-buffer*.

Westover (1989) interprets subvolumes as voxels, but uses a filter kernel to determine their effects on pixels. Data samples are treated as centers of influence, their values assumed to drop off with distance as described by the reconstruction filter. The effect of a data sample with value $\rho(i)$ at location (x, y, z) upon a particular location in space (x_i, y_i, z_i) is found by multiplying the data value by the filter value h_v, which is a function of the distance between them. To find the effect upon a single pixel (the projected location of (x_i, y_i, z_i), the filter must be integrated over the range of z values that will project there. This integral, which is independent of the actual sample value, is referred to as the *footprint(x,y)*. If the three-dimensional filter kernel is rotationally symmetric, the footprint can be precomputed for all viewpoints. If not, it must be computed anew for each viewpoint. In either case, it can be stored in a table. The color and opacity of the voxel are multiplied by the footprint at all pixels that the voxel may effect to give its contribution there.

6.6.4. Compositing: Integrating Through the Volume

Finally, the visual contributions of all the subvolumes must be accumulated, taking into account possible occlusion of some subvolumes by others. To do this physically correctly, one should consider the light scattering (Blinn 1982; Kajiya and Von Herzen 1984; Max 1986; Sabella 1988). However, most volume rendering methods (Drebin et al. 1988; Levoy 1988; Sabella 1988; Upson and Keeler 1988; Westover 1989) use a simplified discrete approach described by Porter and Duff (1984), and this will be considered here. Using methods described in the previous three sections, an intensity and opacity value has been found indicating the contribution of a particular subvolume to a particular pixel. The pixel also has a cumulative intensity value plus a cumulative opacity, which are both initialized to zero.

Using Porter/Duff compositing, the intensity of the composite (I_{comp}) of two contributors is:

$$I_{comp} = I_{front} \times \alpha_{front} + I_{back} \times \alpha_{back} \times (1 - a_{front})$$

(6.2)

Where I_{front} is the intensity of the front (before considering its opacity), I_{back} is the intensity coming from the back, I_{comp} is the intensity of the result. α_{front} is the opacity of the front and α_{back} is the opacity of the back. Sometimes the intensity stored has already taken into account opacity, and it is not multiplied by α during compositing. (Drebin et al. 1988). The opacity of the combination is

$$\alpha_{comp} = \alpha_{front} + (1 - \alpha_{front}) \times \alpha_{back} \qquad (6.3)$$

The total attenuated intensity of a series of regions or surface composited together is then:

$$I_{comp} = \sum_{i=1}^{n} b_i \prod_{j=1}^{i-1} \Theta_j \qquad (6.4)$$

where b_i is the intensity of a region and Θ_j is the attenuation from each region in front of it.

Compositing can be done front-to-back (Upson and Keeler 1988; Westover 1989; Levoy 1989a), in which case the pixel is the front contributor and the subvolume is the back. It can also be done back-to-front (Drebin et al. 1988; Levoy 1988; Westover 1989), in which case the opposite is true. Front-to-back can be more efficient, because the composition can be stopped when the opacity reaches 100%, when nothing further away will be visible. Back-to-front has the advantage that all subvolumes are in front, and potentially viewable, at some point in the imaging process.

6.7. Topics for Future Research

Volume visualization is a young field and much work remains to be done. The more sophisticated approaches cannot be done at interactive speeds because of the often very large data volumes, but ability to interact with the visualization is a major desiderata. Not only is it necessary to view the volume from arbitrary positions to appreciate its three-dimensional complexity, but it is often desirable to interact with the visualization or simulation as it happens to improve the image or steer the computation. Two approaches that will help improve the speed of these approaches are use of parallel processing to allow the simple computational tasks to occur simultaneously, and the use of hierarchical data structures that summarize volume data and help avoid continuous calculations on the voxel level. Another important issue is the generation of the algorithm to visualize irregular, higher-dimensional and multi-parameter data sets.

Acknowledgements

This work has been supported by a California State Microelectronics Grant with matching funds from Digital Equipment Corporation and a UCSC Opportunity Fund Grant. Images were produced on an Iris-4D donated by Silicon Graphics Inc., and on a Titan on loan from Ardent Computers, who also provided the CT skull data. Software for visualization was written by Judy Milton, Madhavan Thirumalai, and Alex Keith, of UCSC.

References

Artzy E, Frieder G, Herman G. The theory, design, implementation, and evaluation of a three-dimensional surface generation program. *Proc. SIGGRAPH, Computer Graphics*, pp.2-9, July 1980.

Artzy E, Frieder G, Herman G. The theory, design, implementation, and evaluation of a three-dimensional surface detection algorithm. *Computer Graphics and Image Processing*, Vol. 15(1), pp.1-24, January 1981.

Bergman L, Fuchs H, Grant E, Spach S. Image rendering by adaptive refinementr. *Proc. SIGGRAPH, Computer Graphics*, August 1986.

Blinn JF. Light reflection functions for simulation for clouds and dusty surfaces. *Proc. SIGGRAPH, Computer Graphics*, July 1982.

Bloomenthal J. Polygonization of implicit surfaces. *Computer-Aided geometric Design*, Vol. 5, pp.341-355, 1988.

Phong BT. Illumination for computer-generated pictures. *Comm. ACM*, pp.311-317, June 1975.

Cline HE, Dumoulin CL, Lorensen WE, Hart Jr. HR, Ludke S. 3D reconstruction of the brain from magnetic resonance images. *Magnetic Resonance Imaging*, July 1987.

Chen LS, Herman GT, Reynolds A, UdupaK. Surface shading in a cuberille environment. *IEEE Computer Graphics and Applications*, Vol. 5(12), pp.33-43, December 1985.

Cook LT, Dwyer III SJ, Batnitzky S, Lee KR. A three-dimensional display system for diagnostic imaging applications. *IEEE Computer Graphics and Applications*, Vol. 3(5), pp.13-19, August 1983.

Cline HE, Lorensen WE, Ludke S, Crawford CR, Teeter BC. Two algorithms for the reconstruction of surfaces from tomograms. *Medical Physics*, June 1988.

Cook R, Porter T, Carpenter L. Distributed ray tracing. *Proc. SIGGRAPH, Computer Graphics*, pp.137-145, 1984.

Christiansen HN, Sederberg TW. Conversion of complex contour line definitions into polygonal element mosaics. *Proc. SIGGRAPH, Computer Graphics*, pp.187-192, August 1978.

Drebin RA, Carpenter L, Hanrahan P. Volume rendering. *Proc. SIGGRAPH, Computer Graphics*, pp.65-74, July 1988.

Dürst MJ. Letters: additional reference to "matching cubes". April 1988.

Farrell E. Color display and interactive interpretation of three-dimensional data. *IBM J. Res. Develop.*, Vol. 27(4), pp.356-366, July 1983.

Frieder G, Gordon D, Reynolds RA. Back-to-front display of voxel-based. *IEEE Computer Graphics and Applications*, Vol. 5(1), pp.52-60, January 1985.

Fuchs H, Kedem ZM, Uselton SP. Optimal surface reconstruction from planar contours. *Comm. ACM*, Vol. 10, pp.693-702, October 1977.

Fujimoto A, Takayuki T, Iwaia K. ARTS: accelerated ray-tracing system. *IEEE Computer Graphics and Applications*, Vol. 6(4), pp.16-26, April 1986.

Foley JD, Van Dam A. *Fundamentals of Interactive Computer Graphics*. Addison-Wesley Publ. Company, Reading, Mass. 1982.

Fowler D, Ware C. Strokes for representing univariate vector field maps. *Proc. Graphics Interface '89*, London, Ontario, pp.249-253, June 1989.

Farrell EJ, Zappulla R, Yang WC. Color 3d imaging of normal and pathologic intracranial structures. *IEEE Computer Graphics and Applications*, Vol. 4(9), pp.5-17, September 1984.

Ganapathy S, Dennehy TG. A new general triangulation method for planar contours. *Proc. SIGGRAPH, Computer Graphics*, pp.69-75, 1982.

Gibson CJ. A new method for the three-dimensional display of tomographic images. *Physics in Medicine and Biology*, Vol. 28(10), pp.1153-1157, 1983.

Gelberg L, Kamina D, Vroom J. Vex: a volume exploratorium. *Proc. Volume Visualization Workshop*, Dept. Computer Science, University of North Carolina, Chapel Hill,NC, pp.21-26, May 1989.

Grinstein G, Pickett RM, Williams MG. Exvis: an exploratory visualization environment. *Proc. Graphics Interface '89*, pp.254-261, June 1989.

Höhne KH, Bernstein R. Shading 3D-images from ct using gray-level gradients. *IEEE Trans. Medical Imaging*, pp.45-57, 1986.

Hu X, et al. Volumetric rendering of multimodality multivariable mediacl imaging data. *Proc. Volume Visualization Workshop*, Dept. Computer Science, University of North Carolina, Chapel Hill, NC, pp.45-49, May 1989.

Helman J, Hesselink L. Representation and display of vector field topology in fluid flow data sets. *Computer*, Vol. 22(8), pp.27, August 1989.

Herman GT, Liu HK. Three-Dimensional display of human organs from computer tomography. *Computer Graphics and Image Processing*, Vol. 9(1) 1979.

Harris LD, Robb RA, Yuen TS, Ritman EL. Proc. SPIE, Vol. 152, pp.10-18, 1978.

Hibbard W, Santek D. Interactivity is the key. *Proc. Volume Visualization Workshop*, Dept. Computer Science, University of North Carolina, Chapel Hill, NC, pp.39-43, May 1989.

Herman GT, Udupa JK. Display of 3-D digital images: computational foundations and mediacl applications. *IEEE Computer Graphics and Applications*, Vol. 3(5), pp.39-46, August 1983.

Keppel E. Approximations of complex surfaces by triangulation of contour lines. *IBM Journal of Research and Development*, Vol. 19, pp.2-11, 1975.

Kajiya J, Von Herzen BP. Ray tracing volume densities. *Proc. SIGGRAPH, Computer Graphics*, pp.165-174, July 1984.

Lorensen WE, Cline HE. Marching cubes: a high resolution 3D surface construction algorithm. *Proc. SIGGRAPH, Computer Graphics*, pp.163-169, July 1987.

Levoy M. Display of surfaces from volume data. *IEEE Computer Graphics and Applications*, Vol. 8(3), pp.29-37, March 1988.

Levoy M. Design for a real-time high-quality volume rendering workstation. *Proc. Volume Visualization Workshop*, Dept. Computer Science, University of North Carolina, Chapel Hill, NC, pp.85-90, May 1989.

Levoy M. Rendering mxtures of geometric and volumetric data (Submitted for publication). 1989.

Liu HK. Two and three-dimensional boundary detection. *Computer Graphics and Image Processing*, Vol. 6(2), pp.134-142, April 1977.

Lee M, Redner R, Uselton S. Statiscally optimised sampling for distributed ray tracing. *Proc. SIGGRAPH, Computer Graphics*, pp.61-67, 1985.

Max N. Light diffusion through clouds and base. *Computer Vision and Image Processing*, Vol. 33, pp.280-292, 1986.

Meagher DJ. Geometric modeling using octree encoding. *Computer Graphics and Image Processing*, Vol. 19, pp.129-147, 1982.

Meagher DJ. Interactive solids processing for medical analysis and planning. *Proc. NCGA 5th Annual Conference*, Dallas, TX, pp.96-106, May 1984.

Meyer G, Greenberg D. Perceptual color spaces for computer graphics. *Proc. SIGGRAPH, Computer Graphics*, pp.247-261, 1980.

Mao X, Kunii TL, Fuhishiro I, Nomoa T. Hierarchical representations of 2D/3D gray-scale images and their 2D/3D two-way conversion. *IEEE Computer Graphics and Applications*, Vol. 7(12), pp.37-44, December 1987.

Norton A. Generation and display of geometric fractals in 3-d. *Proc. SIGGRAPH, Computer Graphics*, pp.3, July 1982.

Porter T, Duff T. Compositing digital images. *Proc. SIGGRAPH, Computer Graphics*, pp.253-260, July 1984.

Peachey DR. Solid texturing of complex surfaces. *Proc. SIGGRAPH, Computer Graphics*, pp.279-286, July 1985.

Pizer SM, et al. 3D shaded graphics in radiotherapy and diagnostic imaging. *Proc. NCGA 86 Conf.*, pp.107-113, 1986.

Press W, Flannery B, Teukolsky S, Vettering W. *Numerical Recipes: The Art of Scientific Computing*. Cambridge University Press 1986.

Pratt W. *Digital Image Processing*. John Wiley and Sons, New York 1978.

Reeves W, Blau R. Approximate and probalistic algorithms for shading and rendering structured particle systems. *Proc. SIGGRAPH, Computer Graphics*, pp.313-322, July 1985.

Reeves W. Particle systems — a technique for modeling a class of fuzzy objects. *Proc. SIGGRAPH, Computer Graphics*, pp.359-373, July 1983.

Robertson PK. The application of scene synthesis techniques to the display of multi-dimensional image data. *ACM Transactions on Graphics*, Vol. 4(4), pp.248-275, October 1975.

Sabin M. A siurvey of contouring methods. *Computer Graphics Forum*, Vol. 5, pp.325-340, 1986.

Sabella P. A rendering algorithm for visualizing 3D scalar fields. *Proc. SIGGRAPH, Computer Graphics*, pp.51-58, July 1988.

Snyder J, Barr A. Ray tracing complex models containing surface tessellations. *Proc. SIGGRAPH, Computer Graphics*, pp.119-128, July 1987.

Sunguroff A, Greenberg D. Computer generated images for medical applications. *Proc. SIGGRAPH, Computer Graphics*, pp.196-202, August 1978.

Stettner A, Greenberg D. Computer Graphics visualization for acoustic simulation. *Proc. SIGGRAPH, Computer Graphics*, pp.195-206, July 1989.

Shirley P, Neeman H. Volume visualization at the center for supercomputing research and development. *Proc. Volume Visualization Workshop*, Dept. Computer Science, University of North Carolina, Chapel Hill, NC, pp.17-20, May 1989.

Srihari SN. Representation of three-dimensional digital images. *Computing Surveys*, Vol. 13(4), pp.399-424, 1981.

Schlusselberg DS, Smith WK, Woodward DJ. Three-dimensional display of medical image volumes. *Proc. NCGA*, pp.114-123, March 1986.

Trousset Y, Schmitt F. Active ray tracing for 3D medical imaging. Proc. Eurographics '87, Amsterdam, pp.139-149, 1987.

Tuy HK, Tuy T. Direct 2-D display of 3-D objects. *IEEE Computer Graphics and Applications*, Vol. 4(10), pp.29-33, October 1984.

Upson C, Keeler M. The v-buffer: visible volume rendering. *Proc. SIGGRAPH, Computer Graphics*, pp.59-64, July 1988.

Upson C. The visual simulation of amorphous phenomena. *The Visual Computer*, Vol. 2, pp.321-326, 1986.

Udupa JK, Srihari SH, Herman GT. Boundary detection in multidimensions. *IEEE transactions on Pattern Analysis and Machine Intelligence*, Vol. 4(1), pp.41-40, January 1982.

Van Gelder A, Wilhelms J. Topological considerations in isosurface generation. 1989, Submitted for publication.

Vannier MW, Marsh JL, Warren JO. Three-dimensional computer graphics for craniofacial surgical planning and evaluation. *Proc. SIGGRAPH, Computer Graphics*, pp.263-273, July 1983.

Westover L. Interactive volume rendering. *Proc. Volume Visualization Workshop*, Dept. Computer Science, University of North Carolina, Chapel Hill, NC, pp.9-16, May 1989.

Wright T, Humbrecht J. Isosurf — an algorithm for plotting isovalued surfaces of a function of three variables. *Proc. SIGGRAPH, Computer Graphics*, pp.182-189, August 1979.

Whitted T. An improved illumination model for shaded display. *Communication of the ACM*, Vol. 23(6), pp.343-349, June 1980.

Wyvill G, McPheeters C, Wyvill B. Data structures for soft objects. *The Visual Computer*, Vol. 2(4), pp.227-234, August 1986.

7

Special Models for Natural Objects

André Gagalowicz
INRIA

This chapter is devoted to special models used to simulate various objects: fractals, particles systems, solid texturing, botanical structures.

7.1. Fractals

Fractal Geometry was introduced by B. Mandelbrot (1983) to deal with complex functions, for instance continuous but nowhere smooth that classical geometry fails to describe. Such functions appear frequently in nature and everyday life, and the usual methods, such as derivation, cannot be used.

7.1.1. Basic idea

The assumption made is that the phenomena studied keep their complexity at each level of analysis. The very well-known example is the measurement of coasts or borderlines between countries. To measure a coast, one can use a rule of length m, and place it along the coast. If the rule has been placed n times, the length L(m) will be: $L(m) = n \times m$. But, as one decreases the length m of the rule, n increases so: $L(m) \to \infty$.

If, instead of using a rule, one tries to measure the length by covering the coast with circles of radius m, and then let m decrease to 0, the same difficulty is encountered. This problem is solved with the notion of fractal dimension. Several definitions exist for it, and the following one corresponds to Haussdorff dimension. To measure a line, lengths are generally added; for surfaces, squares of lengths; for volumes, cubes of lengths, etc.

A shape is said **fractal**, if, to measure it, lengths should be added at a power which is not an integer: for many coasts, there is a number D such that:

$$\lim_{m \to 0} (\sum m^\alpha) = 0 \text{ if } \alpha < D \qquad \lim_{m \to 0} (\sum m^\alpha) = \infty \text{ if } \alpha > D$$

D is the fractal dimension of the shape.

7.1.2. Image synthesis

Many features make fractal sets of great interest for image synthesis; for instance, the property that as the sets are viewed at greater magnification more and more structure is shown, or the fact that only a small number of parameters are needed for their specification, independent of the visual complexity.

Fractals also often provide a convenient way to synthesize some natural objects with significant data compression. Fractal landscapes, clouds or vegetation are quite common.

One problem with fractal techniques is to control precisely the shape of the synthesized object: only a few parameters generally control the fractal dimension, the size and so on.

7.1.3. Iterated Function System

The method of Iterated Function System (IFS), first developed by M. Barnsley (1986), provides a good control and understanding of fractal synthesis.

An IFS is a couple (W, P) where:

. $W = \{w_1, \dots , w_n\}$ is a finite set of mappings, all of them being strict **contractions**.

. $P = \{ p_1, \dots , p_n\}$ is a set of probabilities: $\forall$ i, $0 < p_i \leq 1$, $\Sigma\, p_i = 1$.

$$w_{k(i,j)} = \begin{bmatrix} a_1^k & a_2^k \\ a_3^k & a_4^k \end{bmatrix} \begin{bmatrix} i \\ j \end{bmatrix} + \begin{bmatrix} a_5^k \\ a_6^k \end{bmatrix} \qquad \forall\, k \in \{1, 2, \dots n\} \qquad (7.1)$$

for each k, w_k is such that $\forall\, u \in \mathbf{R}^2$, $|w_k(u)| \leq s_k\, |u|$, $s_k < 1$

We then consider the following process: let z_0 be any point in $\mathbf{R}^2$, we randomly choose a map w_i (with probability p_i), compute $z_1 = w_i\,(z_0)$. We repeat this process a great number of times. If all points are plotted after an great enough number of iterations, they will distribute themselves approximately upon a compact set A, called the **attractor** of the IFS.

With every IFS is associated a unique attractor A, and a special probability measure, the p-balanced measure which is the stationary distribution of the random walk; the probabilities, through this measure, have an influence on the density of the points of A (the grey levels). A characterization of A is given by the following theorem:

if (W, P) is an IFS and A its attractor, then:

$$A = \bigcup_{i=1}^{n} w_i(A)$$

$$(7.2)$$

Moreover, if L is a compact 2-D set of $\mathbf{R}^2$, the "collage theorem" (Barnsley et al. 1986) tells us that if there are n contractions and some $\varepsilon > 0$ such that:

$$h \left(L, \bigcup_{i=1}^{n} w_i(L) \right) < \varepsilon$$

then:

$$h (L, A) < \frac{\varepsilon}{1-s}$$

where A is the attractor of the IFS, h is the Haussdorff metric:

$$h(B, C) = \max_{x \in B} [\max(\min(d(x, y))), \max_{y \in C}(\min(d(x, y)))]$$

for any two subsets B and C, where d is an Euclidian distance and s is such:

$$0 < s < 1 / d(w_i(x), w_i(y)) \leq s\, d(x, y), \qquad \forall \ x, y$$

It follows that to find a suitable set of maps to reconstruct A (or L...), we only need to make an **approximate** covering of L by continuously distorted smaller copies of itself.

The design of fractal objects with IFS is made simple and intuitive with the use of (7.2). This method leads to the inverse problem which can be stated as follows:

7.1.4. Image Analysis

Given a set A, determine an IFS that will approximately generate A.

Some authors have discussed this problem. M. Barnsley has proposed a method based on the moment theory of p-balanced measures that works under certain hypotheses (Barnsley et al. 1986).

The method proposed in (Lévy Véhel and Gagalowicz 1987) has the advantage of being general. It is based upon an optimization technique. Given a 2-D shape A and a value **n** of number of contractions, the problem solved determines the set of parameters $\{a_k^l , k = 1, 2, 3, 4, 5, 6, \ 1 = 1, 2, ... n\}$ which minimizes a criterion related to the error between A and its fractal approximation A_F.

Two criteria have been used:

. **The Haussdorff distance between the two shapes.** As this criterion is highly non differentiable with respect to the contraction parameters, a generalized gradient technique called a bundle algorithm (Lemaréchal et al. 1981) was used to converge to a set of parameters which minimizes **locally** this criterion. If we start too far from the solution, we may thus not converge to an acceptable solution.

. **The simple mean square error between the two images** (which thus allows us to manipulate grey tone 2-D shapes) has the advantage of being differentiable with respect to the parameters. The use of a classical gradient technique allows a more comfortable implementation and convergence.

Lévy Véhel and Gagalowicz (1988) have developed a method based upon the ideas of simulated annealing that allows the generation of a shape A plus the grey levels on A as before, but also insures the convergence to a (the) global minimum.

IFS is a powerful method that makes it possible to synthesize very complex shapes very fast; one can approximate natural objects like the natural maple leaf of figure 7.1 (see colour section), or design complex images (figure 7.2, colour section) with a very few number of parameters: 30 for the maple leaf, 100 for figure 7.2 which gives a data compression ratio of 25000. Figure 7.1 is subdivided into 4 subimages. The top left image presents a natural maple leaf sensed by a black and white camera and slightly colored with a pseudo-color technique. The top right image displays the natural leaf thresholded to isolate it from the background. The two bottom subimages show the fractal approximation obtained of the natural leaf at two different resolutions.

7.2. Particle   Systems

7.2.1. Introduction

A different model applied for the simulation of plants and other objects was proposed by Reeves (1983) who called it: **particle system**. Particle systems have been explored by W.T. Reeves & al. (Reeves 1983; Reeves and Blau 1985) to represent fuzzy objects such as fires, clouds, grass and trees. Objects are represented by a multitude of minute primitives named **particles**. Particle shapes are always very simple, for example: points or spheres can be considered. These particles together represent the volume of an object. At the inverse of classical models, a particle system model is a dynamical model able to represent motion. Over time, particles are generated into the system, move and change state within the system and die from the system.

Particle systems have proven to be powerful modeling techniques for the following reasons.

* A particle is much simpler than most graphical primitives, sophisticated treatments like motion blur can be applied easily.
* A particle model is both a procedural and stochastic model.
* The modeling of quite complex and irregular shapes does not need any sophisticated user specification.
* Movements and distortions of objects can be simulated.
* The resolution of the generated object can be adapted to the viewpoint.

Two drawbacks must also be specified.
* As many particles have to be generated, some specific rendering techniques should be considered
* Mixing particle systems with other scene description is often complex.

Particle systems allow the representation of natural objects as opposed to person-made objects. Dynamical objects such as fires, fireworks, torrents, waves or smoke and static ones such as grass and trees have been modelled. Dynamical phenomena are represented by a sequence of images representing the particle evolution as a function of time. Static objects are often defined by the particle trajectories. One can appraise the realism of particle systems modeling on two well known films. In "Star Trek II: The Wrath of Khan," the Genesis Demo sequence shows the application of particle systems to the representation of fire elements (Paramount 1982). Trees and grass elements were also represented by sophisticated particle systems in the film "The Adventure of André and Wally B." (Lucasfilm 1984).

7.2.2. Models of Particle Systems

In its basic form, a particle system is a collection of particles where each particle is independent of each other. A set of attributes is associated to each particle at its creation. The evolution of a particle system can be described as follows: at each time,

* some particles are added to the system,
* some particles are removed from the system,
* the existing particles are transformed according to their attributes values.

Simple particle attributes can be:

* initial position,
* initial velocity (both speed and direction),
* initial size,
* initial color,
* initial transparency,
* shape,
* lifetime.

The number of particles generated and their associated attributes are computed from some global parameters of the system. These parameters can be distributed among two classes: the **generation parameters** and the **transformation parameters**.

Generation parameters:

Some generation parameters are:
* a position in a three-dimensional space that defines the origin of the particle system,
* two angles of rotation about a coordinate system through origin to define an orientation,
* a generation shape which defines a region about its origin into which newly born particles are randomly placed. Common generation shapes are a sphere, a circle or a rectangle,

* the mean number of new particles *MeanParts$_f$* generated at a frame and its variance *VarParts$_f$* such that the number of particles generated at a frame is:

$$NParts_f = MeanParts_f + Rand() \times VarParts_f,$$

with *Rand* a procedure returning an uniformly distributed random number between -1.0 and +1.0,
* the mean initial speed and its variance, with the same computation as for the number of particles,
* in the same way, an average value and a maximum variation value are also defined for color, transparency, size and lifetime,
* the particle shape.

The system may be also increased by allowing the previous parameters to change over time.

Transformation parameters:

Transformation parameters such as the rate of color change, the rate of transparency change or the rate of size change are also defined.

The simple model presented above allows the representation of dynamical phenomena such as fires or fireworks. Other applications need more structured models. Until now, each particle was independent of each other. An improvement consists in allowing a particle to generate some new particle systems. One can also use a procedural particle system generation process. Trees have been modelled by recursively generating some new particle systems for each sub-branch.

Realistic effects can also be added by allowing particles movement being modified by external phenomena such as wind, pull of gravity etc.

Each application needs the definition of a new set of specific parameters. For example, some specific global parameters used to model trees are the height of the tree, the mean branch length, the width of the tree.

7.2.3. Particle rendering

Let us consider the rendering process associated to a frame. Assuming the position, geometry and appearance parameters of each particle are known, the rendering of particle systems is quite similar to the rendering of common graphical primitives such as polygons or curved surfaces. Unfortunately, the number of particles is so high (usually between 10.000 and 1.000.000) that traditional rendering techniques cannot be applied. This paragraph describes some approximations that can be made to simplify the rendering algorithms. Two steps of the rendering algorithm are considered: the visible surface determination and the shading process.

7.2.3.1. Visible surface determination

The rendering process differs depending on the application. One can distinguish light emitting particles and light reflecting particles. Particle systems simulating fires, fireworks or explosions can be considered in the first class whereas particle systems representing trees or grass lie in the second one.

Light-emitting particles:
Considering particles as light-emitting greatly simplifies the rendering process. Each particle is considered as an independent light source. The display algorithm can be quite simple: each particle adds a bit of light to the pixels that it covers. The amount of light added and its color depend on the particle transparency and color. With this algorithm, no sorting of the particles is needed, the rendering is very efficient.

Light-reflecting particles:
Light-reflecting particles need more sophisticated rendering techniques. For example, algorithms for the display of carpet of grass and forest are presented here. The forest contains many trees. We will first determine in which order trees should be displayed. Displaying trees independently implies that we assume the bounding boxes of the trees do not intersect. Under this assumption, we can sort all the trees into a back-to-front order with respect to screen depth. The algorithm consists in determining the sort list and then for each tree, beginning with the former one, computing and rendering the tree.

The hidden surface determination of each tree is more complex because some branches of a tree can obscure others; a bucket sort is used. The bounding volume of the tree is subdivided into buckets indexed by the eye-space z distance of the particle. Particles are displayed according to this order using the painter's algorithm.

The algorithm chosen to render grass is similar except more accuracy is needed to sort clumps of grass. The non-intersection assumption is not valid any more. Instead of displaying all the buckets associated to a clump at one time, only the buckets that do not intersect the bounding boxes of the remaining clumps in the eye-space z distance are displayed. Remaining particles are then reassigned to the following buckets.

7.2.3.2. Shading

Shading fire or firework particle systems is simple because each particle is independent i.e. cannot play any part in the shading of other particles. On the other hand, motion blur is necessary to solve temporal aliasing and strobing effects. In our application, we can take advantage of the particle simplicity. We can be satisfied with drawing antialiased straight lines instead of points. The straight lines are drawn between the particle positions at the beginning of the frame and halfway through the frame.

Trees and grass particle systems need more sophisticated shading models with ambient, diffuse and specular shading components and shadows. An exact shading is again not possible because of the number of particles in the scene. It is necessary to approximate the shading. The chosen solution consists in using probabilistic shading models. For example, for grass, the contribution of both the diffuse and ambient components depend on the distance from the top of the clump of grass to the particle in question, decreasing exponentially as the depth increases. Similarly, external shadows are added to trees by considering the plane passing through the light source and the top of the neighboring trees. This plane cuts the tree; shadows are defined according to this plane: particles above the plane are in full sunlight and particles below the plane are partially in the shade depending of the distance from the plane to the particle.

7.3. Solid texturing

7.3.1. Introduction

Traditionally, the problem of texturing objects has been solved using surface texturing methods (including the mapping of a 2-D texture on a 3-D object and the procedural generation of a 2-D texture over the surface of a 3-D object). Surface texturing consists of the evaluation of a function T(u,v) defined over a set of points located on the surface of the object being textured.

Solid texturing is the generalization to the 3-D space of the concept of surface texturing. Solid texturing also consists of the evaluation of a space function T(x,y,z) that is defined over the entire volume of the object (often in a cube containing it): it is also defined in the interior points of the object.

A space function can be seen as representing a block of solid material: the evaluation of such a function at the visible points of an object's surface will give the texture of that surface (i.e. the texture that could be seen when sculpturing the first block to get the wanted object). So it is called "solid texture".

Originally, with surface texturing methods, the value returned by the 2-D function was used to determine the reflectance or color at one point of the surface. Any improvement that has been introduced for surface texturing is also available for solid texturing, such as the use of the returned value to control local reflection, shininess, roughness, or surface normal direction. The latter technique allows one to simulate bumps and wrinkles on geometrically smooth surfaces (Blinn 1978; Blinn and Newell 1976).

Solid texturing methods can be used in exactly the same way, so it is possible to simulate anything that can be simulated using surface texturing methods but without the traditional inconveniences of these methods (Perlin 1985).

7.3.2. Solid Texturing Examples

The concept of "solid texture" has been introduced in 1985 independently by Perlin (1985) and Peachey (1985). A wide range of materials and "visual effects" has been simulated using these solid texturing methods. A similar technique has been used by Gardner (1985) to get a "visual simulation of clouds".

1. The 3-D equivalents of the 2-D functions used in previous works for procedural generation of 2-D textures can be used to generate 3-D textures: sine functions, stochastic functions, functional composition.

2. Traditional texture mapping technique can be included in solid texturing by using the composition of a look-up table function (getting values from the texture map) and a projection function from the 3-D space to a 2-D space.
3. The value returned by the space function can be used to control the color (marble texture (Perlin 1985)), intensity or transparency (clouds (Gardner 1985)) of an object.
4. The value returned by the space function can be used to control the direction of the surface's normal vector. This allows the simulation of bump textures (Perlin 1985).
5. The perturbation of the normal direction can be correlated to time: this allows the simulation of waves (Perlin 1985).
6. Both controlling the normal direction and the color as a function of time allows one to simulate fire (Perlin 1985).

7.3.3. Advantages and problems of Solid Texturing

One of the advantages of solid texturing techniques is that texture becomes independent of the object's shape: texture does not have to fit into the object's surface and so we do not have to face the traditional problems encountered when using surface texturing techniques (for example, with the poles of a sphere). A second advantage is that, as for all procedural techniques, the database is extremely small.

Although solid texturing methods are really powerful, there are a certain number of problems that make them not so easy to use. The first problem is that the generated texture depends of the density of sample points on the surface of the objects being textured. The second is there is no way to automatically determine which functions should be used to generate a planned texture: the choice of the texture functions should be mainly based on intuition. The problem is that solid texturing visually simulates the aspect of some materials and natural elements and that this simulation is not based upon reality. So, practice and experience are needed to get convincing results.

7.4. Structural Modeling of Botanical Structures

7.4.1. Introduction

Recently, many attempts have been made to simulate the development of plants, trees and botanical structures. Some interesting results have been obtained using branching process constructions (Ansaldi et al.1985; Aono and Kunii 1984) graftals and particle systems (Reeves 1983; Reeves and Blau 1985; Demko et al. 1985), combinatories on trees (de Does and Lindenmayer 1983). The very first model taking account of the botanical structure of plants and trees has been proposed by de Reffye, Edelin, Françon, Jaeger and Puech (Eyrolles 1986). It is a parametric model which allows the image generation of a wide variety of plants and trees, the integration of time to simulate the growth of plants, and the incidence of wind, insect attacks and other natural parameters. Their system works using the simulation of the activity of **buds** at different times. These buds can either become a flower and die, or go to sleep, or become an internode, or die, etc.

These kinds of events can happen giving specific parameters; each one is part of the model. These parameters are also probabilistic: this explains the wide range of plants that have been rendered using this system.

7.4.2. Simulation of the Architecture of Plants

The model considers a wide variety of structural parameters such as the order of an axis (see figure 7.3), phyllotaxy (see figure 7.4), and ramification (figure 7.5).

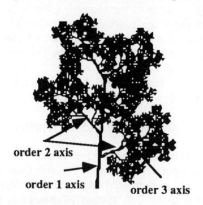

Figure 7.3. The notion of order of an axis

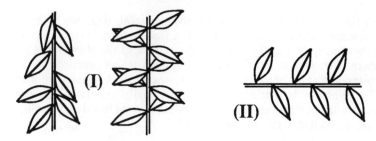

Figure 7.4. Phyllotaxy: (I) spiraled   (II) distic

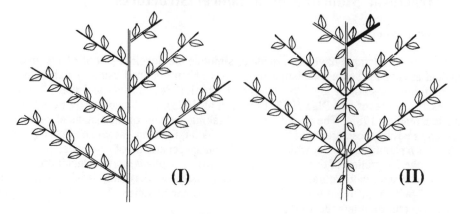

Figure 7.5. Ramification: (I) continuous   (II) rhythmic

The **order of the axis** of a plant is a parameter relating the branching process to the apical buds of a growth unit. **Phyllotaxy** governs the relative position of a lateral bud with the lateral bud of a previous node. **Ramification** can be either continuous, rhythmic or diffuse. It is **continuous** whenever every node of an axis is the root of an axis of greater order, **rhythmic** whenever some nodes are the root of an axis of greater order and **diffuse** whenever these roots are located at random.

Some laws are known between botanical species and these parameters. Hence, as explained in (Eyrolles 1986), the kinds of ramifications are functions of the order of the axis for a given variety. The development trend of a tree may be **orthotropic** or **plagiotropic** (see figure 7.6).

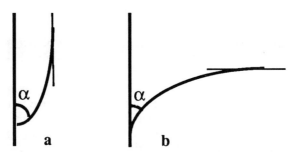

Figure 7.6. Development trend of a tree (left: orthotropy, right: plagiotropy)

Lastly, the model accounts for a tree architecture. Only 23 different architectural models are known (Eyrolles 1986).

The model proposed by these co-authors is a quantitative (but probabilistic) one. Time is digitized so that the basic unit of time is the time taken by the growth of a growth unit, which is supposed to be constant for the axes of a given order (hence axes of different orders do not generally grow at the same speed). To each bud are given two probabilities: an **abortion** probability and a **"wait"** probability (the bud can wait during a certain amount of time without growing).

The ramification process is included in the **"non-abort"** probability parameter. As explained in (Eyrolles 1986), growth simulation is obtained including the following parameters:

- age
- growth of the axes
- number of buds at each node
- probability of death, pause, ramification and reiteration
- type of growth unit
- geometry of internodes
- orthotropy and plagiotropy.

Each of these parameters may correspond to existing ones in nature or not. In the later case, interesting "monsters" may be obtained!

These parameters can produce different types of plants, tree architectures and growing. Some impressive images have been obtained, and other parameters such as bending of branches under the action of gravity or wind have been implemented.

7.4.3. Visualization and rendering

The system produces a set of vectors and faces (usually a great number of them). A classical Z-buffer algorithm is used for the elimination of hidden surfaces. Trunk texture may be obtained using the technique developed by Bloomenthal (1985).

Figure 7.7 (see colour section) displays an image of palm tree built with the use of this model. Figure 7.8 (see colour section) shows a "natural" weeping willow obtained by the same model. Both images are impressive by their realism.

7.5. Free Surface Modeling by Simulated Evolution

7.5.1. Introduction

We describe a procedural method for modeling free surfaces (free in the sense that, for a surface, the exact position of every point and the precise form of the surface about each point is not of primary importance). This method enables the modeling of open or closed planar surfaces, with or without holes, and particularly, the simulation of the evolution of these surfaces. This method (Lienhardt 1987; Lienhardt and Françon 1987) has been applied to the synthesis of both static and dynamic images of natural surfaces, movement of articulated objects, etc.

The method is based on principles common to two methods for the simulation of natural phenomena: the simulation of the growth of plants by internodes developed by P. de Reffye (1979; de Reffye et al. 1988), and particle systems presented by W. Reeves (1983). These principles are:

- **digitizing:** each model is composed of basic elements (particles, internodes) which lead to digitizing in time (denoted h hereafter);

- **notion of element proper activity:** the principle of each method consists in simulating the behavior of basic elements ;

- **characterization of the object evolution** being modelled according to a set of "parameters" associated to the basic elements;

- **priority of topology over geometry,** which assumes increasing importance in three-dimensional modeling (see (Baumgart 1975; Ansaldi et al. 1985; Mantyla 1983; Guibas and Stolfi 1985) for example).

Particle systems are a pointwise model, where the basic element is a point-like particle. de Reffye's model is a linear model where the basic element is a curve-like internode. The method presented here is an extension of the previous methods to the modeling of **surfaces**: it is based upon a surface-like model, called **modular map,** composed with surface elements denoting elementary modules (see figure 7.9). Following a classification proposed by A. Fournier (1987), this model is a **structural** model.

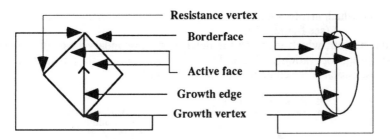

Figure 7.9. Elementary modules (growth edges are represented graphically by directed edges, and resistance edges by undirected edges).

7.5.2. Modular Map

A modular map (see (Tutte 1984; Baumgart 1975; Guibas and Stolfi 1985) for more details) represents a set of faces, edges and vertices modeling the topology of subdivision of a surface. Two kinds of vertices and edges are defined:

- **growth** vertices and edges corresponding to the object skeleton and making up a rooted tree.
- **resistance** vertices and edges, which are passive during the map evolution. They constitute the outline of the object in particular.

Vertices and faces of a modular map are partitioned into growth vertices, **active faces** (incident with at least one growth edge) and resistance vertices, border faces (only incident with resistance edges). Every modular map is either reduced to a vertex which is a growth vertex, or composed of elementary modules. Each elementary module is formed from two distinct triangular active faces, incident to a common growth edge. Each active face is incident with a growth edge and two resistance edges, two growth vertices and a resistance vertex (see figure 7.9).

7.5.3. Evolution of Modular Maps

The method is as follows: the topology of surface S at time h is modelled as a modular map C_h, defined as a set of vertices, edges, and faces. One vertex is distinguished in each modular map and called the **initial vertex**. The modular map C_0, defining the topology of S at h=0 is particularly reduced to the initial vertex. The evolution of the topology of S between times h=0 and h=n is modelled by a sequence of modular maps $(C_0, ..., C_n)$. Each modular map C_i (i >o) is determined from the modular map C_{i-1} by an operation on the map. These operations are:

- the **creation of an initial elementary module** ; the operation is applicable to a modular map reduced to a growth vertex (in particular, this operation applies to the map Co);
- **external ramification**, corresponding to the creation of a new elementary module in the "interior" of a border face (see figure 7.10);
- **internal ramification** or the creation of a new elementary module in the "interior" of an activity face (see figure 7.11);
- **deletion** of an elementary module;

. **fusion**, which involves the creation of a border face defined by two resistance edges (see figure 7.12);

. **disjunction**, which involves the deletion of a boundary face (the inverse of the preceding operation). These last two operations, in particular, permit the representation of the contour of the modelled surface.

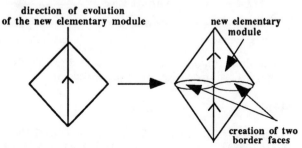

Figure 7.10. External ramification of a modular map

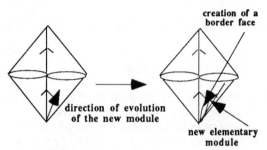

Figure 7.11. Internal ramification

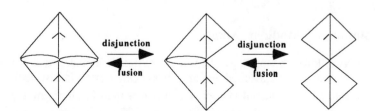

Figure 7.12. Fusion / disjunction of a modular map

The principle of the characterization of a sequence of modular maps is inspired by graph **grammars** that is, the characterization needs the definition of:

. a **finite alphabet** (set of "labels"),

. an **initial modular map** (C_0, reduced to a vertex),

. **"production rules"** composed of a left-hand side, which is a label, and a right-hand side defining a sequence of operations applicable to a modular map.

If several sequences of operations are applicable to a single label, it is possible to have conditions which specify processing (cf. figure 7.13). These conditions use parameters associated to edges and to the initial vertex of the map (these parameters are detailed below).

The principle is as follows. A label is associated to every edge and to the initial vertex. Whenever this label appears as the left-hand side of a production rule, the sequence of operations defineu by the right-hand side is applied (i.e. to every edge and to the initial vertex). This is repeated at each time h for every edge or initial vertex.

7.5.4. Characterization of Modular Maps

Before describing the parameters associated to the edges and to the initial vertex of a modular map, it must be pointed out that the growth elements (vertices and edges) form a tree, called the growth tree. This tree is directed: the initial vertex of the map is the root of the tree, and the tree is directed from the root to the leaves. The growth tree of a modular map is structured along axes; that is: an integer, called the order of the edge, is associated with each edge. An axis is a maximal length path in the growth tree (given the directions of the growth edges), with all the edges of the path having the same order.

Various parameters may be associated to the edges and to the initial vertex of a modular map which permit the characterization of applicable operations.

Given an edge A and the growth edge A' such that both are incident with the same elementary module, then associated with A are (cf. figure 7.13):

- the time value h;
- the label of A;
- the order of A;
- the order of A';
- the time of creation of A, each edge (and each vertex, excepting the initial vertex of the map) being created at time h> 0;
- the type of A, depending on whether A is incident with the root or with a leaf of the growth tree;
- distance relationships between A' and some other element X of the growth tree; the distance between A' and X is the number of edges in the unique, simple elementary path between them (for example, the distance from A' to the root of the growth tree, or the origin of the axis incident with A').

The submersion of the surface in the usual three-dimension space, and the aspect (limited to color) are defined at each instant by real functions associated with the vertices and faces of the modular map (defining the positions of the vertices and faces of the modular map and the colors of its faces). These real functions are first defined by the user in the production rules. The functions take as arguments the characteristics defined above (time, date of creation, etc...).

The surface modeling method defined here allows us to simulate the evolution of the topology of the surfaces (for growth simulation, for example), and the evolution of purely geometric properties (that is, the positions of the vertices of the map which allows us to simulate the evolution of articulated objects), and the coloring of the surface (for example, the color variation of a leaf in autumn).

Figure 7.14 (see colour section) presents the evolution of a maple leaf using this technique and figure 7.15 (see colour section), a beautiful bell flower.

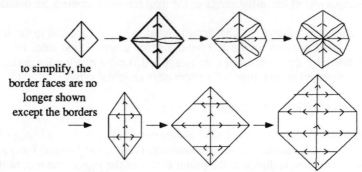

to simplify, the border faces are no longer shown except the borders

Figure 7.13. An example of evolution.

The sequence of modular maps is obtained by successive applications of external and internal ramifications. The growth tree is composed of two types of axes: an order 1 axis (vertical), and order 2 axes (horizontal). Each clock time produces the creation of a new elementary module whose growth edge extends the axis of order 1, and the creation of elementary modules whose growth edges extend an already existing axis of order 2, or is the origin of a new axis of order 2. Modules whose growth edge extend an axis of order 2 are only created if the length of the axis (number of growth vertices forming the axis) is strictly less than the distance (the number of growth vertices) between the origin of the secondary axis and the root of the tree which accounts for the halt in growth of the secondary inferior axes.

Acknowledgments

I would like to thank L. Bourcier, S. Coquillart, F. Cros, J. Lévy-Véhel, P. Lienhardt, P. Sander and H. Yahia who helped me in the preparation of this chapter.

References

Ansaldi S, de Floriani L, Faldieno B, (1985) Geometric Modeling of Solid Objects by Using a Face Adjacency Graph Representation; *Proc. SIGGRAPH '85, Computer Graphics*, Vol.19, N.3, pp. 131-139.

Aono M and Kunii TL (1984) Botanical Tree Image Generation, *IEEE Computer Graphics and Applications*, vol. 4, N°5, pp. 10-23.

Barnsley M (1986) *Making Chaotic Dynamical Systems to Order*, lecture notes, INRIA Workshop on Fractals.

Barnsley M, Ervin V, Hardin D and Lancaster J, (1986) Solution of an Inverse Problem for Fractals and Other Sets, *Proc. Nat. Acad. Sci.*, USA, VOL 83.

Baumgart B, (1975) A Polyhedron Representation for Computer Vision , *AFIPS*, Proc. 44, pp. 589-596.

Blinn JF (1978) Simulation of Wrinkled Surfaces. *Proc. SIGGRAPH '78, Computer Graphics*, vol.12 , N.3, pp. 286-292.

Blinn JF and Newell ME (1976) Texture and Reflexion in Computer Generated Images. *Comm. ACM*, vol.10, N.10, pp.542-547.

Bloomenthal J, (1985) Modeling the Mighty Maple, *Proc. SIGGRAPH '85, Computer Graphics*, Vol. 19, N.3, pp. 305-311.

de Reffye P, (1979) Modelisation de l'Architecture des Arbres par des Processus Stochastiques: Simulation Spatiale des Modèles Tropicaux sous l'Effet de la Pesanteur. Application au Coffea Robusta, Thèse de doctorat d'Etat, Université de Paris Sud, Orsay.

de Reffye P, Edelin C, Françon J, Jaeger M, Puech C (1988) Plant Models Faithful to Botanical Structure and Development, *Proc. ACM Siggraph'88*, Atlanta.

de Does M, Lindenmayer A, (1983) Algorithms for the Generation and Draving of Maps Representing Cell Clones, Lectures Notes in Computer Science, Vol. 15, N. 3, pp. 301-316.

Demko S, Hodges L, Naylor B, (1985) Construction of Fractal Objects with Iterated Function Systems, *Proc. SIGGRAPH '85, Computer Graphics*, Vol. 19, N.3, 1985, pp. 271-278.

Eyrolles G, (1986) Synthèse d'Images Figuratives d'Arbres par des Méthodes Combinatoires, Thèse de troisième cycle, Université de Bordeaux 1, 1986.

Eyrolles G, Françon J, Viennot G (1986) Combinatoire pour la Synthèse d'Images Réalistes de Plantes, *Actes du deuxiemème colloque Image*, CESTA, pp. 648-652.

Fournier A, (1987) "The Modeling of Natural Phenomena", Course n°16, Siggraph'87.

Gardner GY (1985) Visual Simulation of Clouds. *Proc. SIGGRAPH '85, Computer Graphics*, vol.19, N.3, pp.297-303.

Guibas L, Stolfi J, (1985) Primitives for the Manipulation of General Subdivisions and the Computation of Voronoi Diagrams, *A.C.M. Transactions on Graphics*, Vol. 4, N.2, pp. 74-123.

Hallé F, Oldeman R, Tomlison P, (1978) *Tropical Trees and Forest: an Architectural Analysis*, Springer-Verlag.

Kawagachi Y (1982) A Morphological Study of the Form of Nature *Computer Graphics*, vol. 16, N.3, pp. 223-232.

Lienhardt P, (1987) Modelisation et Evolution de Surfaces Libres, Thèse de Doctorat, Université Louis Pasteur, N.196.

Lienhardt P, Françon J, (1987) Vegetal Leaves Images Synthesis MARI 87, 3e semaine de l'Image Electronique, Paris (La vilette), 18-22 May 1987

Lemaréchal C, Strodiot JJ, Bihian A (1981) On a bundle algorithm for nonsmooth optimization, *Non-Linear Programming 4*, Mangasarian, Meyer, Robinson Editors (Academic Press), pp. 245-282.

Lévy Véhel J and Gagalowicz A (1987) Shape Approximation by a Fractal Model, *Proc. EUROGRAPHICS'87*.

Lévy Véhel J and Gagalowicz A (1988) Shape Approximation of 2-D Objects, *Proc.EUROGRAPHICS'88*.

Lucasfilm Ltd, (1984) *The Adventures of André and Wally B.*, (film), August 1984.

Magnenat-Thalmann N, Thalmann D, (1985) Three-Dimensional Computer Animation: More an Evolution Than a Motion Problem, *IEEE Computer Graphics and Applications*, pp. 47-57.

Mandelbrot B(1983) *The Fractal Geometry of Nature* , W. H. Freeman and Co., San Francisco.

Mantyla M, (1983) Computational Topology: a Study of Topological Manipulations and Interrogations in Computer Graphics and Geometric Modeling, *Acta Polytechnica Scandinava*, N.37.

Oppenheiner PE, (1986) Real Time Design and Animation of Fractal Plants and Trees, *Proc.SIGGRAPH '86, Computer Graphics*, vol. 20, N.4, pp. 55-64.

Paramount, Genesis Demo from Star Trek II (1982): The Wrath of Khan, in *Siggraph Video Review* Number 11, June 1982.

Peachey DR (1985) Solid Texturing of Complex Surfaces. *Proc. SIGGRAPH '85, Computer Graphics*, vol.19, N.3, pp.279-286.

Perlin K (1985) An Image Synthesizer. *Proc. SIGGRAPH '85, Computer Graphics*, vol.19, N.3, pp.287-296.

Reeves W, (1983) Particle System: a Technique for Modeling a Class af Fuzzy Objects, *A.C.M. Transactions on Graphics*, Vol.2, N.2, pp. 91-108.

Reeves, WT, Blau R, (1985) Approximate and Probabilistic Algorithms For Shading and Rendering Structured Particle Systems, *Computer Graphics*, vol. 19, N.3, pp. 313-322.

Smith AR, (1984) Plants, Fractals and Formal Languages, *Proc. SIGGRAPH '84, Computer Graphics*, vol.18, N.3, pp. 1-10.

Stevens P, (1974) *Patterns in Nature*, Little Brown and Co.

Tutte W, (1984) *Graph Theory* , Encyclopedia of Mathematics and Its Applications, Vol. 21, Addison-Wesley.

8

Computer Animation: a Tool that Gives Life to Visualization

Nadia Magnenat Thalmann
University of Geneva

8.1. Introduction

Visualization has become an important way of validating new models created by scientists. When the model evolves over time, computer simulation is generally used to obtain the evolution of time, and computer animation is a natural way of visualizing the results obtained from the simulation. By using computer animation, scientists may better understand how various phenomena evolve in space and in time. In the future, the application of computer animation to the scientific world will become very common in many scientific areas: fluid dynamics, molecular dynamics, thermodynamics, plasma physics, astrophysics etc. In this chapter, we discuss the basic techniques of computer animation, which may be applied to the visualization of most models evolving over time. Note however that computer animation will be used to simulate the impact of new technologies on human beings, and animation sequences will more and more often involve human characters. Consequently, research in animation of articulated bodies is also essential and will be also discussed in this chapter.

8.2. What is Computer Animation ?

8.2.1. The term "Computer Animation"

Most of the phenomena which may be represented on the screen of a workstation are typically time-dependent. The techniques of computer graphics allow the construction of 3D graphical objects using geometric modeling techniques. Moreover, in a 3D space, scenes are viewed using virtual cameras and they may be lit by synthetic light sources. In order to visualize these phenomena at any given time, it is necessary to know the appearance of the scene at this time and then Computer Graphics techniques allow us to build and display the scene according to viewing and lighting parameters. The problems to solve are how to express time dependence in the scene, and how to make it evolve over time.

These problems and their various solutions are part of what is usually understood by the term Computer Animation (Magnenat-Thalmann and Thalmann 1985).

The term "Computer Animation" suggests that computers bring something new to the traditional way of animating. Traditional animation is defined as a technique in which the illusion of movement is created by photographing a series of individual drawings on successive frames of film. Is this definition, due to John Halas (Halas and Manwell 1968), still true for Computer Animation ? The definition is essentially correct if we change the definition of the words *photographing*, *drawings*, and *successive frames*. A definition of computer animation could be: a technique in which the illusion of movement is created by displaying on a screen, or recording on a recording device a series of individual states of a dynamic scene. We use the term "recording" also for photographing and we consider both a cine-camera and a videorecorder as a recording device.

Our definition corresponds to the various techniques for producing films using a computer: computer-aided animation, real-time animation, procedural animation, and automatic motion control.

8.2.2. Traditional animation and computer animation

Traditional animated films are produced as a sequence of images recorded frame-by-frame. Computer-animated films may be produced using the exact same process. What is different in this case is the way the frames were produced. In a traditional film, frames are drawings created by artists. In a computer-animated film, frames are produced by the computer based on the artist's directives. As we shall discuss later on, directives could vary from directly drawing each frame on a graphics tablet to just giving orders to three-dimensional characters. The results will be the same: a series of images produced by the computer.

8.2.3. Real-time animation and frame-by-frame animation

One interesting question is how long it takes for the computer to produce a frame. The answer is easy; from almost no time at all to an unlimited time depending on the complexity of the images and the computer used. With powerful graphics workstations, reasonably complex scenes may be rendered in a fraction of second. For example, a typical workstation (e.g. Silicon Graphics IRIS 4D) has the theoretical capabilities to render 60000 polygons in one second. Now consider one second of animation: it needs at least 15 frames to represent a relatively smooth animation, because the illusion of continuous movement breaks down at slower speeds. More typically, animators use 24 frames/sec for films, and 25 or 30 frames/sec for video (depending on the country). For a 25 frames/sec film produced on our workstation, we may theoretically use 2400 (60000/25) polygons per frame. Of course, this is not really possible, because of the time necessary to compute the animation. What is important is that it is possible to see the animation directly on the workstation screen, without recording it. Such real-time animation is more and more popular. In the future, we may expect that workstations will be able to produce more complex images in 1/25 of a second. However, the complexity of images is also increasing very fast. This means that real-time animation and frame-by-frame animation will always exist. However the level of complexity dividing the two types of animation will be much higher in the future.

8.3. Keyframe animation

8.3.1. Introduction to keyframe animation

A computer animated sequence is obtained by a series of images produced by the computer according to the animator's directives. We may distinguish three general methodologies:

1. All frames are given to the computer by the animator. This means for example that the animator will draw all 2D drawings on a graphics tablet using a paint system. The role of the computer is almost nonexistent. Another approach, called **rotoscopy**, consists of recording the motion by a specific device for each frame and using this information to generate the image by computer. For example, a human walking motion may be recorded and then applied to a computer-generated 3D character. This approach will provide a very good motion, because it comes directly from reality. However, it does not bring any new concept to animation methodology, and for any new motion, it is necessary to record the reality again.

2. The second and most used method is called **keyframe animation**. It consists mainly of giving to the computer a certain number of frames, called **keyframes**, and the computer derives the other frames using interpolation procedures. Various approaches have been developed in this context. The two most important are shape interpolation and parametric keyframe animation. In shape transformations, the "inbetweens" are obtained by interpolating the keyframe images themselves. In parametric keyframe animation, objects are characterized by parameters; motion is specified by giving key values for each parameter; inbetween values are calculated using an interpolation law.

3. In the third approach, called **procedural animation**, motion is algorithmically described.

8.3.2. Shape interpolation

In two dimensions, the computer is generally used to help the production of animation, what is called **computer-assisted animation**. In this area, shape interpolation has been popular, especially because of the artist Peter Foldes, who created marvellous films based on this technique: *Hunger* (1974) and *Metadata* (1971). This is an old technique, introduced by Burtnyk and Wein (1971). Each keyframe is characterized by key points which have to correspond. When corresponding keyframes do not have the same number of key points, it is necessary to add extra-key points as shown in Fig. 8.1.

This method may be extended to three-dimensional objects. The principle is the same when objects are modeled in wire-frame. However, the technique is much more complex when objects are facet-based, because a correspondence between facets and between vertices must be found. Vertices and facets must be added in order to have the same numbers for both objects. A complete algorithm has been introduced by Hong et al. (1988).

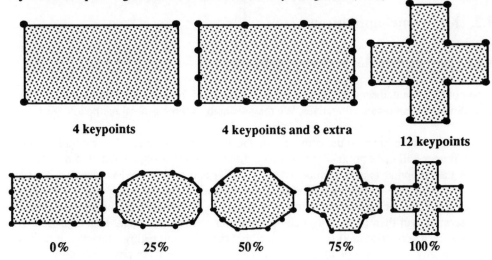

4 keypoints   **4 keypoints and 8 extra**

12 keypoints

0%   25%   50%   75%   100%

Figure 8.1. An example of interpolation with two different number of points

8.3.3. The problem of interpolation

Once the correspondence has been found, the problem is to interpolate corresponding elements. The general problem of interpolation may be summarized as follows: given a certain number of points, find a curve passing through these points. The simplest method consists of joining the points by straight lines; this is a linear interpolation:

$$\text{Position}(t) = (1\text{-}t)\, \text{Position}_i + t\, \text{Position}_j \qquad (8.1)$$

However, if this method is used for animation, it causes a lack of smoothness which considerably alters the motion. Undesirable effects are produced such as discontinuities in the speed of motion and distortions in rotations, as shown in Fig. 8.2. Consequently, for such motions, others methods should be used. A non-linear expression may be used instead of Equation (8.1); however, this is a non-linearity in time and not in space.

Two interesting methods for solving this problem are the moving point constraint method (Reeves 1981) and the Kochanek-Bartels (1984) spline interpolation.

The Reeves method is meant to allow the specification of multiple paths and speeds of interpolation and to reduce motion discontinuities at key frames. The principle of the technique is to associate a curve varying in space and time with points of the animated object. This curve is called **moving point** and it controls the trajectory and the dynamics of the point. Similar to a P-curve (Baecker 1969), the moving point has the shape of the trajectory, but a trail of symbols is used instead of a continuous line to depict the path. The symbols are spaced equally in time. This means that the dynamics are represented by the local density of the symbols.

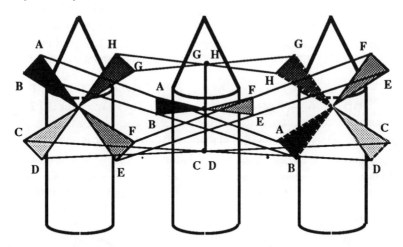

Figure 8.2. In this example, linear interpolation produces undesirable distortions

The Kochanek-Bartels method of interpolating splines is based on piecewise continuous interpolation of the curve by cubic functions. The interpolating curve must be continuous at the given points only up to a certain order. The method allows us to control the curve at each given point by three parameters: tension, continuity and bias. It works as follows. Consider a list of points P_i and the parameter t along the spline to be determined. A point V is obtained from each value of t from only the two nearest given points along the curve (one behind P_i, one in front of P_{i+1}). But, the tangent vectors D_i and D_{i+1} at these two points are also necessary. This means that, we have:

$$V = THC^T \qquad (8.2)$$

where T is the matrix $[t^3\ t^2\ t\ 1]$, H is the Hermite matrix, and C is the matrix $[P_i, P_{i+1}, D_i, D_{i+1}]$. The Hermite matrix is given by:

$$H = \begin{bmatrix} 2 & -2 & 1 & 1 \\ -3 & 3 & -2 & -1 \\ 0 & 0 & 1 & 0 \\ 1 & 0 & 0 & 0 \end{bmatrix} \qquad (8.3)$$

Kochanek and Bartels consider the source derivative (tangent vector) DS_i and the destination derivative (tangent vector) DD_i at any point P_i. Using these derivatives, they propose the use of three parameters to control the splines—**tension, continuity,** and **bias.** The tension parameter t controls how sharply the curve bends at a point P_i. As shown in Fig. 8.3 a-b, in certain cases a wider, more exaggerated curve may be desired, while in other cases the desired path may be much tighter. The continuity c of the spline at a point P_i is controlled by the parameter c. Continuity in the direction and speed of motion is not always desirable. Animating a bouncing ball, for example, requires the introduction of a discontinuity in the motion of the point of impact, as shown in Fig. 8.3 c-d. The direction of the path as it passes through a point P_i is controlled by the bias parameter b. This feature allows the animator to anticipate a trajectory or overshoot a key position by a certain amount, as shown in Fig. 8.3e. Equations combining the three parameters may be obtained:

$$DS_i = \; 0.5 \; [(1-t)(1+c)(1-b) \; (P_{i+1}-P_i) + (1-t)(1-c)(1+b) \; (P_i-P_{i-1})] \quad (8.4)$$
$$DD_i = \; 0.5 \; [(1-t)(1-c)(1-b) \; (P_{i+1}-P_i) + (1-t)(1+c)(1+b) \; (P_i-P_{i-1})] \quad (8.5)$$

A spline is then generated using Eq. (1) with DD_i and DS_{i+1} instead of D_i and D_{i+1}. The technique may be used for shape interpolation but it is widely used in parametric keyframe animation: interpolation of angles in the animation of human bodies, interpolation of facial parameters, interpolation of control points for the design of camera and light paths.

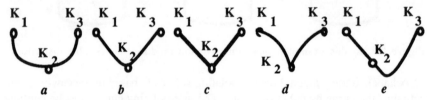

Figure   8.3. *a-b.* Variation of tension: the interpolation in *b* is more tense than the interpolation in *a; c-d.* Variation of continuity: the interpolation in *b* is more discontinuous than the interpolation in *a; e.* A biased interpolation at K_2

8.3.4. Parametric key-frame animation

Parametric key-frame animation is based on the following principle: an entity (object, camera, light) is characterized by parameters. The animator creates keyframes by specifying the appropriate set of parameter values at a given time, parameters are then interpolated and images are finally individually constructed from the interpolated parameters. Linear interpolation causes first-derivative discontinuities, causing discontinuities in speed and consequently jerky animation. The use of high-level interpolation such as the Kochanek-Bartels spline interpolation is preferable. Fig. 8.4 (see color section) shows an example of parametric keyframe animation.

8.4. Procedural animation

In **procedural animation**, motion is algorithmically described. Let us have an example to understand the concept and compare it with keyframe animation. We assume an object falling from a height h=100. We would like to generate the animation sequence corresponding to the motion of this object. We know from physics that the equation of motion is $y = h - 0.5g \; t^2$.

With an approximate value of the acceleration gravity g, we have the law $y = 100 - 5 \, t^2$. It means that after 1 second, the object will be at the height $y_1 = 95$, after 2 sec. at the height $y_2 = 80$, after 3 sec. at the height $y_3 = 55$, after 4 sec. at the height $y_4 = 20$. At the time $t=5$, the object is already on the floor. A typical keyframe animation consists of giving to the computer the following set: {<0,100>, <1,95>, <2,80>, <3,55>, <4,20>} and to expect that the computer will generate frames for each 1/25 sec. By a linear interpolation, it will be unrealistic. Using a quadratic interpolation, the result will be exact, because the physics law is a quadratic law. However, it is more direct to use the quadratic equation as follows:

```
e.g.   create OBJECT (...);
       TIME = 0;
       while  Y > 0 do
          Y = 100 - 5*TIME^2
          MOVE_ABS (OBJECT, X,Y,Z);
          draw OBJECT;
          record OBJECT
          erase OBJECT
          TIME:=TIME+1/25;
```

The advantages of using algorithmic (e.g. physical) laws for animation is that it will give the exact animation. It is in fact a pure simulation. Physical laws are applied to parameters of the animated objects (e.g. positions, angles). Control of these laws may be given by programming as in ASAS (Reynolds 1982) and MIRA (Magnenat-Thalmann and Thalmann 1983) or using an interactive director-oriented approach as in the MIRANIM (Magnenat-Thalmann et al. 1985) system. With such an approach, any kind of law may be applied to the parameters. For example, the variation of a joint angle may be controlled by kinematic laws as well as dynamic laws.

To improve computer animation, attention needs to be devoted to the design of evolution laws (Magnenat-Thalmann and Thalmann 1985b). Animators must be able to apply any evolution law to the state variables which drive animation. These evolution laws should not be restricted to analytical laws. Many complex motions are exactly described by simultaneous differential equations. An analytical solution is generally impossible to obtain. Instead of approximating the motion by a series of rotations and translations, which may be difficult to determine, it is better to use the original equations and a continuous simulation approach. Many examples of animation sequences may be generated using this approach: e.g. rabbit-dog problem, camera in a golf ball, orbiting satellite. How can new laws of evolution based on differential equations be introduced into an animation system? It is essential that any new evolution law be defined in such a way that it may be applied to any state variable. The following strategy (Magnenat-Thalmann and Thalmann 1986b) should be applied:

1. Initialization of global state variables in the animation system.

2. Definition of a procedure P_f, called at each frame, which must be responsible for the integration of the equations and for calculating the values of state variables at the current time.

3. Definition of evolution laws which depend on the global state variables. They compute values that depend on the value of the state variables which are accessed by the procedure P_f during the animation process.

With this strategy, an evolution law may be completely changed at any time (and consequently at any frame).

Unfortunately, there are several drawbacks to this kind of animation:

. It could be very difficult to find the laws corresponding to a given motion.
. When several motions are joined, like a linear accelerated motion followed by a circular motion, animation could be jerky at the transition.
. Animation systems based on physical laws provide a limited number of laws unless there is a mechanism of interpretation of mathematical formula.

Shape interpolation, parametric keyframe animation and procedural animation may be described in a more general and unified way. An animated object is characterized by a set of state variables that drive its motion. The evolution of the state variables is defined by an evolution law. The three types of animation may be redefined using the following terminology as shown in Fig. 8.5.

TYPE OF ANIMATION	STATE VARIABLES	EVOLUTION LAWS
shape interpolation	vertices	linear interpolation spline interpolation Reeves interpolation
parametric keyframe animation	parameters	linear interpolation spline interpolation
procedural animation	parameters	physical laws

Figure 8.5. State variables and evolution laws

8.5. Choreography in 3D animation

8.5.1. Organization of a scene

An animation sequence is composed of one or several related scenes and is characterized by a description, called a **script**. Each scene contains static objects grouped into a **decor**. Movable objects, called **actors** change over time, by **transformations** and **movements** defined by the animator. These transformations are controlled by **laws** that are applied to **variables** which drive the transformations. The transformations may be very simple, like rotations, or very complex, including torsions, flexion, or deformations. Decors and actors are colored and lit by **light sources**. Finally, decors and actors are viewed using **virtual cameras**. These cameras may evolve over time as though manipulated by cameramen.

In order to create all the entities and motions, coordinate and synchronize them, known collectively as **choreography**, the director should use an animation system.

8.5.2. Decors and actors

As in a theater, a decor is a collection of static objects. For example, a room may be a decor for a scene. It may be composed of a table, chairs, a carpet and a lamp. Even in the absence of actors, an animated scene may be produced by moving a camera or changing a light parameter over time. Typically, an animator defines a decor as a list of existing objects in his/her database.

Actors are not necessarily human; a film may involve animals or any object which changes over time. For example, a clock is a non-human actor. Its animation consists of moving the clock pendulum using the corresponding physical law. More generally, we shall define an animated object that changes over time according to a list of transformations.

8.5.3. Camera animation

A scene is only meaningful when it is viewed. But there are many ways a scene may be viewed. It depends on the position of the viewer, where the view is directed, and the viewing angle. Such characteristics, and others, are generally grouped into an entity called a **synthetic** or **virtual camera**. A basic synthetic camera is characterized by at least two parameters: the eye and the interest point. The eye is a point and it represents the location of the camera; the interest point is the point towards which the camera is directed. A viewing angle may also be defined for controlling how wide the observer view is.

One of the most impressive effects in computer-generated films is the possibility of rotating around a three-dimensional object or entering inside any complex solid. Although people generally find these effects very spectacular, they are in fact quite easy to produce. Typical effects used by cameramen may also be simulated as shown in Fig. 8.6.

Panning Effect	The camera is moved horizontally from one point to another one.
Tilting Effect	The camera is moved vertically from one point to another.
Tracking Effect	The camera is moved towards the interest point.
Zooming Effect	A zoom lens permits the cameraman to quickly adjust the size of the field being filmed by the camera. In a virtual camera, a zoom may be defined by changing the ratio between the dimensions in the image space and the display space.

Figure 8.6. Camera effects

The use of several cameras allows the simulation of special effects (Magnenat-Thalmann and Thalmann 1986) like those traditionally produced by optical printers. An optical printer is a movie-camera which is focused on the gate of a lensless movie projector to duplicate one piece of film onto another. Optical printers are used for superimpositions and multiple-image effects. They are also very useful for providing fade and wipe effects. A fade-in is an effect used at the beginning of a scene: the scene gradually appears from black. A fade-out is an effect used at the end of a scene; the scene gradually darkens to black. Such fade effects as well as cross-dissolve effects may be produced by applying evolution laws to the parameters which characterized a color filter of the virtual camera. With the wipe effect, one scene appears to slide over the preceding scene.

8.5.4. Camera path

Although classical camera motions are often used, there are many situations where it may be more attractive to design a nonlinear trajectory for the camera. We call such a trajectory a **camera path**. In fact, a path may be defined for each camera attribute and there are several ways of defining such a path:

1. An analytical law is defined for the eye or the interest point. For example, the camera eye follows a uniform circular motion law.

2. The eye or the interest point is assigned to a moving object. When the actor is moving, the motion of the camera has a similar motion.

3. An interactive curve is built and it is used as a path containing time and space information like a P-curve. Curves may be improved or interactively modified using techniques such as splines. For example, Fig. 8.7 shows the steps to create a spline for moving the eye camera.

Step 1	Positioning of the camera by deciding its initial characteristics: eye, interest point and view angle
Step 2	Definition of the camera eye as the first control point of the spline
Step 3	Other control points are created by moving the camera eye and inserting them as control point for the spline
Step 4	Each control point may be modified and new control points may be inserted between existing control points
Step 5	A time may be defined at each control point
Step 6	All inbetween points should be displayed for control purposes
Step 7	Spline parameters (bias, tension, continuity) are defined at each control point
Step 8	The spline is created

Figure 8.7. Creation steps for a camera spline motion

8.5.5. Animation of Lights

A scene should also receive light to be realistic. **Synthetic lights** should be created; their characteristics may also vary over time. In particular, intensities and positions of source lights may change according to evolution laws. Light motion may also be defined using either parametric keyframe animation or algorithmic animation. Characteristics of source lights (intensity, direction, position, concentration) may be constant or defined as variables. When they are variables, any evolution law may be applied to these variables. For example, we may define a positional light source and assume that the light intensity varies from red to green and the location changes according to an oscillation movement.

8.6. Animation of Synthetic Actors

In three-dimensional character animation, the complexity of motion may be arbitrarily divided into three parts: body animation, hand animation, facial animation.

8.6.1. Skeleton and body animation

When the animator specifies the animation sequence, he/she defines the motion using a skeleton. A skeleton (Badler and Smoliar 1979) is a connected set of segments, corresponding to limbs, and joints, as shown in Fig. 8.8. A joint is the intersection of two segments, which means it is a skeleton point where the limb which is linked to the point may move. The angle between the two segments is called the joint angle. A joint may have at most three kinds of position angles: flexion, pivot and twisting. The **flexion** is a rotation of the limb which is influenced by the joint and cause the motion of all limbs linked to this joint. This flexion is made relatively to the joint point and a flexion axis which has to be defined. The **pivot** makes the flexion axis rotate around the limb which is influenced by the joint. The **twisting** causes a torsion of the limb which is influenced by the joint. The direction of the twisting axis is found similarly to the direction of the pivot.

The mapping of surfaces onto the skeleton may be based on various techniques. Chadwick et al. [1989] propose an approach where the control points of geometric modeling deformations are constrained by an underlying articulated robotics skeleton. Komatsu [1988] describes the synthesis and the transformation of a new human skin model using the Bézier surfaces. Magnenat-Thalmann and Thalmann (1987) introduced the concept of Joint-dependent Local Deformation (JLD) operators, which are specific local deformation operators depending on the nature of the joints. These JLD operators control the evolution of surfaces and may be considered as operators on these surfaces. Each JLD operator will be applicable to some uniquely defined part of the surface which may be called the domain of the operator. The value of the operator itself will be determined as a function of the angular values of the specific set of joints defining the operator. Fig. 8.9 (see color section) shows an example of surface mapping based on JLD operators.

Name	Number	Angles
VERTEBRA 1	2	FTP
VERTEBRA 2	3	FTP
VERTEBRA 3	4	FTP
VERTEBRA 4	5	FTP
VERTEBRA 5	6	FTP
LEFT CLAVICLE	7	FP
RIGHT CLAVICLE	11	FP
LEFT SHOULDER	8	FTP
RIGHT SHOULDER	12	FTP
LEFT ELBOW	9	FT
RIGHT ELBOW	13	FT
LEFT WRIST	10	FP
RIGHT WRIST	14	FP
LEFT HIP	15	F
RIGHT HIP	20	F
LEFT THIGH	16	FTP
RIGHT THIGH	21	FTP
LEFT KNEE	17	F
RIGHT KNEE	22	F
LEFT ANKLE	18	F
RIGHT ANKLE	23	F
LEFT TOE	19	F
RIGHT TOE	24	F

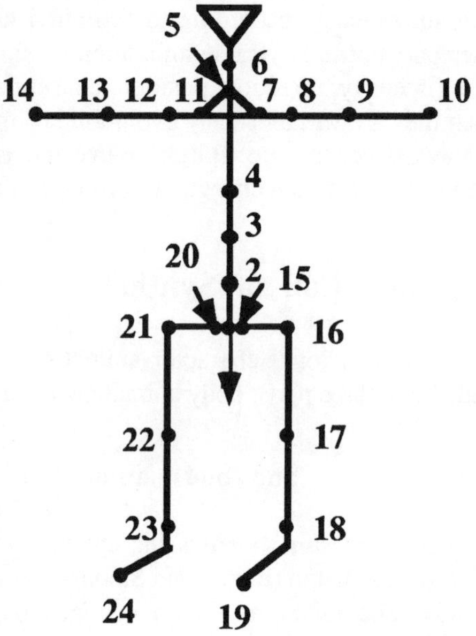

Fig. 8.8. A basic skeleton with the joints angles (F: flexion, T: twisting, P:pivot)

Skeleton animation consists in animating joint angles. According to section 8.3, there are two main ways to do that: parametric keyframe animation and procedural animation based on mechanical laws. In the first method, the animator creates keyframes by specifying the appropriate set of parameter values; parameters are interpolated and images are finally individually constructed from the interpolated parameters. For example, to bend an arm, it is necessary to enter into the computer the elbow angle at different selected times. Then the software is able to find any angle at any time using for example interpolating splines. In the second approach, angles are calculated by applying forces and torques to the limbs in order to vary the angles. This latter approach has been recently introduced by several authors (Armstrong and Green 1985; Wilhelms and Barsky 1985; Girard and Maciejewski 1985: Isaacs and Cohen 1987; Arnaldi et al. 1989).

8.6.2. Hand animation and object grasping

Very few scientific papers have been dedicated to hand animation: in the MOP system designed by Catmull (1972), hands are decomposed into polygons, but undesired variation of finger thickness may occur. Badler and Morris (1982) propose a model based on B-spline surfaces. The surface is computed based on the skeleton of the hand; spheres are linked to the B-spline surface, which does not itself appear on the image. It is only used for the placement of the spheres.

As for the body, two problems are important in hand animation: skeleton motion and surface deformation. Moreover, a connected problem is the problem of grasping an object. When an object is grasped and moved to a new position, it is the arm which essentially guides the object. However to grasp the object, flexion angles have to be determined and this problem is known in robotics as an inverse kinematics problem.

Magnenat-Thalmann et al. (1988a) describe algorithms and methods used to animate a hand for a synthetic actor. The algorithms not only allow the hand to move and grasp objects, but also they compute the deformations of the hands: rounding at joints and muscle inflations. Fig. 8.10 (see color section) shows an example. Gourret et al. [1989] propose a finite element method (see chapter 5) to model the deformations of human flesh due to flexion of members and/or contact with objects.

8.6.3. Facial animation

One of the ultimate objectives is to model exactly the human facial anatomy and movements which satisfy both structural and functional aspects of simulation. This however, involves many problems to be solved simultaneously. Some of these are: the geometric representation must be very close to the actual facial structure and shape, modeling of interior facial details such as muscles, bones, tissues, and incorporating the dynamics and movements involved in making expressions etc. Each one of these is, in itself, a potential area of research.

This complexity leads to what is commonly called facial expressions. The properties of these facial expressions have been studied for 25 years by Psychologist Ekman, who proposed a parameterization of muscles with their relationships to emotions: the Facial Action Coding System (FACS) (Ekman and Friesen, 1978). FACS describes the set of all possible basic actions performable on a human face. Various facial animation approaches have been proposed: parameterized models, muscle model for facial expressions, abstract muscle action procedures, interactive simulation system for human expressions.

A set of parameters is defined based on the desired expressions on a given face. In parameterized models, as introduced by Parke (1982), there are two types of parameters considered: parameters for controlling conformation and parameters controlling expressions. Conformation parameters pertain to the set of parameters which vary from individual to individual and make each person unique. For example, skin color, neck length and shape, shape of chin, forehead, nose length etc. The development of a truly complete set for conformation parameters is very difficult. Since most expressions relate to eyes and mouth for facial expression, primary importance was given to these areas to form a set of expression parameters including pupil dilatation, eyelid opening, eyebrow position, jaw rotation.

Platt and Badler (1981) have presented a model of a human face and developed a notational system to encode actions performable on a face. The notation drives a model of the underlying muscle structure which in turn determines the facial expression. Two separate notations for body and action representation are given: action based notation and structure based notation. Several action based notation systems are explored and ultimately FACS is considered for their model. At the level of structure-based notation, the action representations are interpreted and the simulation of action is performed. The representation adopted, simulates points on the skin, muscle, and bone by a set of interconnected 3D network of points using arcs between selected points to signify relations.

Waters (1987) proposes a muscle model which is not specific to facial topology and is more general for modifying the primary facial expression. The muscle actions are tested against FACS which employs action units directly to one muscle or a small group of muscles. The muscle model was designed to be as naturalistic as possible in the representation. The model is based on a few dynamic parameters that mimic the primary biomechanical characteristics (Waters 1989). Muscles can be described with direction and magnitude. The direction is towards a point of attachment on the bone, and the magnitude of the displacement depends upon the muscle spring constant and the tension created by a muscle contraction. The main parameters employed are: the vector contraction, the position of the head and tail of the vector and fall-off radius. Three types of muscles are created using the muscle model: linear/parallel muscles that pull, sphincter muscles that squeeze and sheet muscle that behaves as a series of linear muscles.

Magnenat Thalmann et al. (1988) introduced a new way of controlling the human face and synchronizing speech based on the concept of abstract muscle action (AMA) procedures. An AMA procedure is a specialized procedure which simulates the specific action of a face muscle. These AMA procedures are specific to the simulated action and are not independent i.e. these are order dependent. These procedures work on certain regions of the human face which must be defined when the face is constructed. Each AMA procedure is responsible for a facial parameter corresponding approximately to a muscle, for example, vertical jaw, close upper lip, close lower lip, lip raiser etc. The complex motion is produced by an AMA procedure. AMA procedures are considered as the basic level, there are two more higher levels defined in order to improve the user interface. These are expression level and script level. At the expression level, a facial expression is considered as a group of facial parameter values obtained by the AMA procedures in different ways. Phonemes and emotions are the two types of facial expression considered. Fig. 8.11 (see color section) shows an example.

Guenter [1989] proposes an interactive system for attaching muscles and wrinkle lines to arbitrary rectangular face meshes and then for simulating the contractions of those muscles. The system is divided into two modules: the Expression Editor and the Muscle and Wrinkler Editor. The Expression Editor uses the FACS to control facial expression at a very high level, The Muscle and Wrinkler Editor is used to established the correspondence between the high level FACS encoding and low level muscular action of the face. The skin is modeled as a linear elastic mesh with zero initial tension and the muscles are modeled as force vectors applied at the vertices of the face mesh.

References

Armstrong WW and Green MW (1985) Dynamics for Animation of Characters with Deformable Surfaces in: N.Magnenat-Thalmann and D.Thalmann (Eds) *Computer-generated Images*, Springer, pp.209-229.

Arnaldi B., Dumont G., Hégron G., Magnenat-Thalmann N., Thalmann D. (1989) Animation Control with Dynamics in: *State-of-the-Art in Computer Animation*, Springer, Tokyo, pp.113-124

Badler NI and Morris MA (1982) Modelling Flexible Articulated Objects, *Proc. Computer Graphics '82*, Online Conf., pp.305-314.

Badler NI and Smoliar SW (1979) Digital Representation of Human Movement, ACM Computing Surveys, March issue, pp.19-38.

Baecker R (1969) Picture-driven Animation, *Proc. Spring Joint Computer Conference*, AFIPS Press, Vol.34, pp.273-288.

Burtnyk N, Wein M (1971) Computer-generated Key-frame Animation, *Journal of SMPTE*, 80, pp.149-153.

Catmull E (1972) A System for Computed-generated movies, *Proc. ACM Annual Conference*, pp.422-431.

Chadwick J, Haumann DR, Parent RE (1989) Layered Construction for Deformable Animated Characters, *Proc. SIGGRAPH '89, Computer Graphics*, Vol. 23, No3, pp.234-243

Ekman P and Friesen W (1978) Facial Action Coding System, Consulting Psychologists Press, Palo Alto.

Girard M and Maciejewski AA (1985) Computational Modeling for Computer Generation of Legged Figures, *Proc. SIGGRAPH '85, Computer Graphics*, Vol. 19, No3, pp.263-270

Gourret JP, Magnenat-Thalmann N, Thalmann D (1989) Simulation of Object and Human Skin Deformations in a Grasping Task, *Proc. SIGGRAPH '89, Computer Graphics*, Vol.23, Vol.4, pp.21-30.

Guenter B (1989) A System for Simulating Human Facial Expression, in: Magnenat-Thalmann N, Thalmann D (Eds) *State-of-the-Art in Computer Animation*, Springer, Tokyo, pp. 191-202

Halas J, Manwell R (1968) *The Technique of Film Animation*, Hastings House, New York

Hong TM, Magnenat-Thalmann N, Thalmann D (1988) A General Algorithm For 3-D Shape Interpolation In a Facet-based Representation, *Proc. Graphics Interface '88*, Edmonton, Canada

Isaacs PM and Cohen MF (1987) Controlling Dynamic Simulation with Kinematic Constraints, Bahavior Functions and Inverse Dynamics, *Proc. SIGGRAPH'87, Computer Graphics*, Vol.21, No4, pp.215-224

Kochanek D and Bartels R (1984) Interpolating Splines with Local Tension, Continuity and Bias Tension, *Proc. SIGGRAPH '84, Computer Graphics*, Vol.18, No3, pp.33-41.

Komatsu K (1988) Human Skin Model Capable of Natural Shape Variation, *The Visual Computer*, Vol.3, No5, pp.265-271

Magnenat-Thalmann N, Laperrière R and Thalmann D (1988a) Joint-dependent Local Deformations for Hand Animation and Object Grasping, *Proc. Graphics Interface '88*

Magnenat-Thalmann N, Primeau E, Thalmann D (1988) Abstract Muscle Action Procedures for Human Face Animation, *The Visual Computer*, Vol.3, No5

Magnenat-Thalmann N, Thalmann D (1983) The Use of High Level Graphical Types in the MIRA Animation System, *IEEE Computer Graphics and Applications*, Vol. 3, No 9, pp. 9-16.

Magnenat-Thalmann N, Thalmann D (1985) *Computer Animation: Theory and Practice*, Springer, Tokyo

Magnenat-Thalmann N, Thalmann D (1985b) 3D Computer Animation: More an Evolution Problem than a Motion Problem, *IEEE Computer Graphics and Applications* Vol. 5, No10, pp. 47-57

Magnenat-Thalmann N, Thalmann D (1986) Special cinematographic effects using multiple virtual movie cameras, *IEEE Computer Graphics and Applications* 6(4):43-50

Magnenat-Thalmann N, Thalmann D (1986b) Three-dimensional computer animation based on simultaneous differential equations, *Proc. Conf. Continuous Simulation Languages*, San Diego, Society for Computer Simulation, pp.73-77

Magnenat-Thalmann N, Thalmann D (1987) The direction of synthetic actors in the film Rendez-vous à Montréal, *IEEE Computer Graphics and Applications*, Vol.7, No12, pp.9-19

Magnenat-Thalmann N, Thalmann D, Fortin M (1985) MIRANIM: an extensible director-oriented system for the animation of realistic images, *IEEE Computer Graphics and Applications*, Vol.5, No3, pp. 61-73

Parke FI (1982) Parameterized Models for Facial Animation, *IEEE Computer Graphics and Applications*, Vol.2, No 9, pp.61-68

Platt S, Badler N (1981) Animating Facial Expressions, *Proc. SIGGRAPH '81, Computer Graphics*, Vol.15, No3, pp.245-252.

Reeves WT (1981) Inbetweening for computer animation utilizing moving point constraints. *Proc. SIGGRAPH '81, Computer Graphics*, Vol.15, No3, pp.263-269

Reynolds CW (1982) Computer Animation with Scripts and Actors, *Proc. SIGGRAPH'82, Computer Graphics*, Vol 16, No3, pp.289-296.

Waters K (1987) A Muscle Model for Animating Three-Dimensional Facial Expression, *Proc. SIGGRAPH '87, Computer Graphics*, Vol.21, No4, pp.17-24.

Waters K (1989) Modeling 3D Facial Expressions: Tutorial Notes. part 2: A Dynamic Model of Facial Tissue, *SIGGRAPH '89 Course Notes on Facial Animation*, pp.145-152.

Wilhelms J and Barsky B (1985) Using Dynamic Analysis to Animate Articulated Bodies such as Humans and Robots, in: N.Magnenat-Thalmann and D.Thalmann (Eds) *Computer-generated Images*, Springer, pp.209-229.

9

Robotic Methods for Task-level and Behavioral Animation

Daniel Thalmann
Swiss Federal Institute of Technology, Lausanne

9.1. Introduction

One of the most important categories of figures in Computer Animation is the articulated figure. There are three ways of animating these linked figures:

1. by recreating the tools used by traditional animators
2. by simulating the physical laws which govern motion in the real world
3. by simulating the behavioral laws which govern the interaction between the objects.

The first approach corresponds to methods heavily relied upon by the animator: rotoscopy, shape transformation, parametric keyframe animation (see Chapter 8). The second way guarantees a realistic motion by using kinematics and dynamics. The problem with this type of animation is controlling the motion produced. The third type of animation is called behavioral animation and takes into account the relationship between each object and the other objects. Moreover the control of animation may be performed at a task- level.

In summary, at a time t, the methodology of calculating the 3D scene is as follows:

1. for a keyframe system: by interpolating the values of parameters at given key times
2. in a dynamic-based system: by calculating the positions from the motion equations obtained with forces and torques
3. for a behavioral animation system: by automatic planning of the motion of an object based on information about the environment (decor and other objects).

Although the problems to solve are not exactly the same, there is a lot to learn in human animation from robot motion. In this chapter, we try to summarize robotics techniques which are used or could be used for the task-level animation of articulated bodies. Simulation of behavior is also briefly discussed.

9.2. Task-level Animation

As stated by Zeltzer (1985), a task-level animation system must schedule the execution of motor programs to control characters, and the motor program themselves must generate the necessary pose vectors. To do this, a knowledge base of objects and figures in the environment is necessary, containing information about their position, physical attributes, and functionality. With task-level control, the animator can only specify the broad outlines of a particular movement and the animation system fills in the details. Task-level motor control is a problem under study by roboticists.

In a robot task-level system (Lozano-Perez 1982), a task planner would transform the task-level specifications into manipulator-level specifications. In a task-level animation system, task-level commands should be transformed into low-level instructions such as a script for algorithmic animation or key values in a parametric keyframe approach. Zeltzer (1982) outlined one approach to task-level animation in which motor behaviour is generated by traversing a hierarchy of skills, represented as frames (Minsky 1975) or actors (Hewitt 1977) in object-oriented systems. Task planning may be divided into three phases:

1) **World modelling**: it consists mainly of describing the geometry and the physical characteristics of the objects and the object. The legal motions of the objects depend on constraints due to the presence of other actors in the environment. The form of the constraints depends itself on the shape of the objects, which requires geometric descriptions of all elements. Another important aspect in task planning is based on the limits of the primitive capabilities of the objet e.g. joint limits for an articulated actor.

2) **Task specification**: there are three ways of specifying tasks in a task-level system:
   1. by example, a method suitable in robotics, but not in animation
   2. by a sequence of model states: each state is given by the configuration of all the objects in the environment. The configuration may be described by a set of spatial relationships as described by Popplestone et al. (1980)
   3. by a sequence of commands; typical examples of commands for synthetic actors (Magnenat-Thalmann and Thalmann 1987) are: WALK to <location>, PICK UP <object> and MOVE it to <location>

3) **Code Generation**: in robotics, the output of the synthesis phase is a program in a manipulator-level language. In computer animation, several kinds of output code are possible: series of frames ready to be recorded, value of parameters for certain keyframes, script in an animation language (Reynolds 1982, Magnenat-Thalmann and Thalmann 1983), script in a command-driven animation system (Magnenat-Thalmann et al. 1985)

9.2.1. Walking

To generate the motion corresponding to the task "WALK from A to B", it is necessary to take into account the possible obstacles, the nature of the terrain and then evaluate the trajectories which consist of a sequence of positions, velocities and accelerations. Given such a trajectory, as well as the forces to be exerted at end effectors, it is possible to determine the torques to be exerted at the joints by inverse dynamics and finally the values of joint angles may be derived for any time. In summary, the task-level system should integrate the following elements: obstacle avoidance, locomotion on rough terrains, trajectory planning, kinematics and dynamics. Using such an approach, however all people with the same physical characteristics would walk exactly the same way which is completely unrealistic. A behavioral approach is necessary, because walking is influenced by the psychology of the individuals. Therefore the animator should be able, for instance, to generate motion corresponding to the task "WALK from A to B in a happy way".

Description of biped gait, and especially human gait, may be easily found in the robotic literature especially concerning biomechanics for constructing artificial walking systems (Gurfinkel and Fomin 1974, Hemami and Farnsworth 1977, Miura and Shimoyama 1984; McMahon 1984, Raibert and Sutherland 1983).

One of the most complete descriptions of the motions of the limbs during walking is due to Saunders et al. (1953) reformulated by McMahon (1984). This can be applied to computer-generated walking. In this description, six models were proposed with an increasing complexity relative to the joints involved.

1. In the first model, called the compass gait, only flexion and extensions of the hips are permitted. The stance leg remains stiff and the trunk moves in an arc with each step.
2. In the second model, the pelvis may turn about a vertical axis; for normal walking, the rotation angle varies between -3° and 3°. This new degree-of-freedom is responsible for a longer step length and a greater radius for the arcs of the hip, hence a smoother ride.
3. The third model adds a pelvic tilt: the pelvis is lowered on the swing-leg side just before toe-off of the swing leg; then it is raised slowly until heel-strike. In this model, the arcs specifying the trajectory of the center of the pelvis are made flatter, because the hip on the swing side falls lower than the hip on the stance side.
4. In the fourth model, the knee flexion of the stance leg is added, which further flattens the arcs specifying the trajectory of the center of the pelvis.
5. The fifth model adds a plantar flexion of the stance; most of this flexion occurs just before toe-off.
6. The sixth model adds a lateral sine motion of the pelvis; the frequency of this motion is half the frequency of the up-and-down motions. It adds a lot of realism to walking appearance, because it simulates the transfer of the weight bearing alternatively from one limb to the other, and the rocking of the body from side to side.

McMahon (1984) proposes a ballistic walking model consisting of three links, one for the stance leg and one for the thigh and shank of the swing leg. The foot of the swing leg is attached rigidly to the shank at right angles. The mass of the trunk, arms and head is lumped at the hip. The mass of the legs is assumed to be distributed over their length so that the center of mass of the thigh is C_1 from the hip, and the center of mass of the shank, including the foot, is C_2 from the knee. McMahon also assumes that the toe of the swing leg does not strike the ground during midswing. The results of the model may be expressed as the normalized time of swing T_s as a function of normalized step length s_l.

$$T_s = \frac{T}{T_n} \qquad T_n = \pi \sqrt{\frac{I}{mgL}} \qquad (9.1)$$

where T is the swing time in seconds, T_n is the natural half-period of the leg as a rigid pendulum, I is the moment of inertia of the rigid leg about the hip, and L is the distance of the leg's center of mass from the hip.

At the behavioral level, we will have to take into account the group level behavior as well as the individual behavior. Everybody walks more or less the same way, following more or less the same laws. This is the "more or less" which is difficult to model. And also a person does not always walk the same way everyday. If the person is tired, or happy, or just got some good news, the way of walking appears slightly different. So in the future, another big challenge is open for the computer animation field: to model human behavior taking into account social differences and individualities. In this direction, Boulic et al. (1990) presents a human walking model built from experimental data based on a wide range of normalized velocities. The model is structured on two levels. At a first level, global spatial and temporal characteristics are generated. At the second level, a set of parameterized trajectories produce both the position of the body in the space and the internal body configuration in particular the pelvis and the legs. The model is based on a simple kinematic approach designed to keep the intrinsic dynamic characteristics of the experimental model. Such an approach also allows a personification of the walking action in an interactive real-time context in most cases.

9.2.2. Grasping

To generate the motion corresponding to the task "PICK UP the object A and PUT it on the object B", the planner must choose where to grasp A so that no collisions will result when grasping or moving it (this is the reach problem). Then grasp configurations should be chosen so that the grasped object is stable in the hand (or at least seems to be stable); moreover contact between the hand and the object should be as natural as possible. Once the object is grasped, the system should generate the motions that will achieve the desired goal of the operation. A free motion should be synthesized; during this motion the principal goal is to reach the destination without collision, which implies obstacle avoidance. In this complex process, joint evolution is determined by kinematics and dynamics equations. In summary, the task-level system should integrate the following elements: path planning, obstacle avoidance, stability and contact determination, kinematics and dynamics. Psychological aspects are also very important and the subtleties of motion should be dependent on the character's personality (Cohen 1989). Fig. 9.1 (see color section) shows an example of a grasping sequence.

The reach problem (Korein and Badler 1982) is the first important problem to be solved for object grasping. As for manipulators in robotics, we may define a workspace for a synthetic actor. Such a workspace is the volume of space which the end-effector of the actor can reach. In a task environment, the reach problem may only be solved if the specified goal point lies within the workspace. Two types of workspace are generally defined:

1. the dextrous workspace, corresponding to a workspace which the actor can reach with all orientations.
2. the reachable workspace, corresponding to a workspace which the actor can reach in at least one orientation.

Consider, for example, a very simple arm with two links and two joints. As shown in Fig. 9.2, the link lengths are L_1 and L_2. We assume that both joint angles α_1 and α_2 may vary between 0^0 and 360^0. Two cases may be considered:

1^0 $L_1=L_2$ The dextrous workspace corresponds to the point S and the reachable workspace is a disc of radius $2L_1$.

2^0 $L_1 \neq L_2$ There is no dextrous workspace and the reachable workspace is a ring of inner radius $|L_1-L_2|$ and outer radius L_1+L_2. Fig. 9.3 shows the workspace.

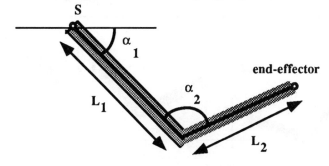

Figure 9.2. A simple 2D arm

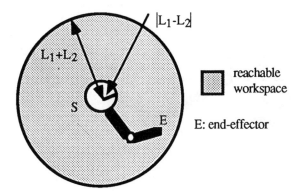

Figure 9.3. 2D reachable workspace

The capability for an articulated body to attain general goal positions and orientations is dependent on the number N of degrees of freedom of the body. When N is less than 6, articulated bodies cannot reach general goals. For example, the 2D arm in Fig. 9.2 cannot reach out of the plane. A typical synthetic actor has about 50 degrees of freedom; however, joint angles have limit values, and it is difficult to represent the reachable workspace.

The second problem is the multiplicity of solutions. How to select the right one ? For example consider the 2D arm of Fig. 9.4a and the goal G. As shown in Fig. 9.4b-c, there are two solutions. For a 2D 3-joint arm, there are at most 4 solutions. In the general case, with N degrees of freedom, the number of solutions is at most 2^N. However, the exact number of solutions depends on the link parameters. For revolute joints:

· the lengths l_k (shortest distances measured along the common normal between the joint axes)
· the twist angles α_k (angles between the axes in a plane perpendicular to l_k)
· the distances d_k between the normals along the joint axes
· the allowable range of motion of the joints (restrictions on angle θ_k)

Moreover, a few solutions may be impossible because of obstacles, as shown in Fig.9.5.

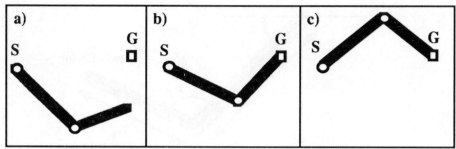

Figure 9.4. **a.** initial conditions **b.** first solution **c.** second solution

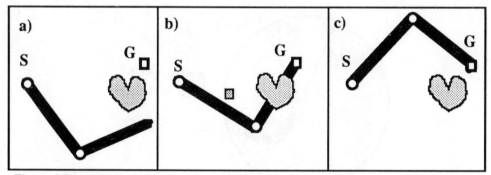

Figure 9.5. **a.** initial conditions **b.** impossible because of the obstacle **c.** unique solution

9.3. Kinematics

9.3.1. A few definitions

In robotics, only rigid bodies are considered; they represent mechanism links. These links are interconnected by joints. Eleven types of joints may be considered between two links. We only mention two of them: revolute joints corresponding to rotational motions, sliding joints corresponding to translational motions. Typically, flexions (e.g. knee and elbow flexions) and twists (e.g. pronation of the forearm) are revolute joints. Other types of joints may be expressed as a combination of revolute and sliding joints. To work with vectors, coordinate systems (called frames) should be selected; the most convenient choice of frames uses axes that move along with the links themselves.

9.3.2. The direct-kinematics problem

This problem consists in finding the position and orientation of a manipulator (the external coordinates) with respect to a fixed-reference coordinate system as a function of time without regard to the forces or the moments that cause the motion. In summary, solving the direct-kinematics problem corresponds to solving the equation $\mathbf{R} = f(\theta)$ where $\mathbf{R} = (r_x,$

$r_y, r_z, r_\rho, r_\theta, r_\psi)$ is the position of the end effector in Cartesian coordinates and $\theta = (\theta_1, \theta_2, \theta_3, \theta_4, \theta_5, \theta_6)$ is the vector of joint angles.

Efficient and numerically well-behaved methods exist for the transformation of position and velocity from joint-space to Cartesian coordinates. To explain the problem, we present the classical example of a manipulator consisting of 6 links and 6 joints, using the Denavit-Hartenberg method (1955). In this method, a coordinate system (body-attached frame) is attached to each joint.

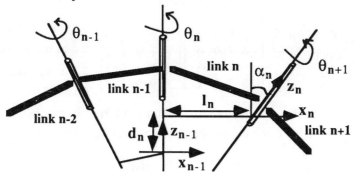

Figure 9.6. Geometry of links and joints

As shown in Fig. 9.6, each link k is characterized by four geometric quantities:

1. the length l_k which is the shortest distance measured along the common normal between the joint axes.
2. the twist angle α_k which is defined as the angle between the axes in a plane perpendicular to l_k.
3. the distance d_k between the normals along the joint axis.
4. the joint angle θ_k.

For a revolute joint **of a robot**, the three quantities l_k, α_k and d_k are constants and θ_k is the only joint variable. For a prismatic joint **of a robot**, the three quantities l_k, α_k and θ_k are constants and d_k is the only joint variable. We assume the following conventions:

- the z_{k-1} axis lies along the axis of motion of the k-th joint
- the x_k axis is normal to the z_{k-1} axis, pointing away from it
- the y_k axis is chosen in order to have a right-handed coordinate system

The relationship between successive frames (k-1,k) is defined by the four transformations:

1. rotation $R_{z\theta}$ of an angle θ_k about the z_{k-1} axis in order to align the x_{k-1} axis with the x_k axis.
2. translation T_{zd} along the z_{k-1} axis of a distance d_k to bring the x_{k-1} axis and the x_k axis into coincidence.

3. translation T_{xl} along the x_k axis of a distance l_k to bring the two origins into coincidence.
4. rotation $R_{x\alpha}$ of an angle α_k about the x_k axis in order to bring to bring the two coordinate systems into coincidence.

The product of the four homogeneous matrices provides the Denavit-Hartenberg matrix:

$$A_k = R_{z\theta} T_{zd} T_{xl} R_{x\alpha} =$$

$$\begin{bmatrix} \cos\theta_k & -\sin\theta_k & 0 & 0 \\ \sin\theta_k & \cos\theta_k & 0 & 0 \\ 0 & 0 & 1 & 0 \\ 0 & 0 & 0 & 1 \end{bmatrix} \begin{bmatrix} 1 & 0 & 0 & 0 \\ 0 & 1 & 0 & 0 \\ 0 & 0 & 1 & d_k \\ 0 & 0 & 0 & 1 \end{bmatrix} \begin{bmatrix} 1 & 0 & 0 & l_k \\ 0 & 1 & 0 & 0 \\ 0 & 0 & 1 & 0 \\ 0 & 0 & 0 & 1 \end{bmatrix} \begin{bmatrix} 1 & 0 & 0 & 0 \\ 0 & \cos\alpha_k & -\sin\alpha_k & 0 \\ 0 & \sin\alpha_k & \cos\alpha_k & 0 \\ 0 & 0 & 0 & 1 \end{bmatrix}$$

$$= \begin{bmatrix} \cos\theta_k & -\sin\alpha_k\cos\theta_k & \sin\alpha_k\sin\theta_k & l_k\cos\theta_k \\ \sin\theta_k & \cos\alpha_k\cos\theta_k & -\sin\alpha_k\cos\theta_k & l_k\sin\theta_k \\ 0 & \sin\alpha_k & \cos\alpha_k & d_k \\ 0 & 0 & 0 & 1 \end{bmatrix} \qquad (9.2)$$

The A_k matrix relates positions and directions in link k to link k-1:

$$^{k-1}x = A_k \, ^k x \qquad (9.3)$$

The position and the orientation of the manipulator end-effector W_6 is the matrix product:

$$W_6 = A_1 A_2 A_3 A_4 A_5 A_6 \qquad (9.4)$$

Fig. 9.7 (see color section) shows an example of a robotic arm manipulated by direct kinematics using the FIFTH DIMENSION Toolkit (Turner et al. 1990).

9.3.3. The inverse-kinematics problem

The inverse-kinematics problem involves the determination of the joint variables given the position and the orientation of the end of the manipulator with respect to the reference coordinate system. This is the key problem, because independent variables in a robot as well as in a synthetic actor are joint variables. Unfortunately, the transformation of position from Cartesian to joint coordinates generally does not have a closed-form solution. However, there are a number of special arrangements of the joint axes for which closed-form solutions do exist, e.g., manipulators having 6 joints, where the three joints nearest the end effector are all revolute and their axes intersect at a point: the wrist. The problem has been studied by numerous authors, but the first efficient algorithm was described by Featherstone (1983) and Hollerbach and Sahar (1983).

The steps of such an algorithm, based on Fig. 9.8, are as follows:

Step 1: Find the wrist position $\vec{p_4}$

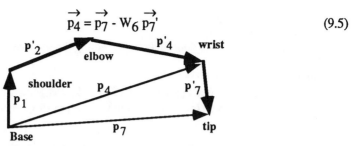

$$\vec{p_4} = \vec{p_7} - W_6 \vec{p_7}'$$ (9.5)

Fig. 9.8. Manipulator geometry

T_6 is the hand orientation matrix, $\vec{p_7}$ the position of the tip of the hand and $\vec{p_7}'$ is the internal vector in link 6 that extends from the coordinate 6 origin to the hand tip.

Step 2: Find the first three angles
The first angle is directly found from the previous equation and the two other angles are easily found if $\vec{p_4}$ is expressed in joint 1 coordinates; because it reduces the problem to a planar two-link problem.

Step 3: Find the orientation relative to the forearm
The forearm orientation is given by $W_3 = A_1 A_2 A_3$ so that the hand orientation relative to the forearm is given by:

$$^3W_6 = A_3^T A_6$$ (9.6)

Step 4: Find the last three joint angles
Joint angles θ_4, θ_5 and θ_6 are found from the elements of the matrix $^3W_6$, which are identified by some matrix manipulations, described by Hollerbach and Sahar (1983).

In a typical system based on inverse kinematics, the animator specifies discrete positions and motions for end parts; then the system computes the necessary joint angles and orientations for other parts of the body to put the specified parts in the desired positions and through the desired motions. Such an approach works well for simple linkages. However, the inverse kinematic solutions to a particular position become numerous and complicated, when the number of linkages increases.

9.4. Dynamics

9.4.1. Methods based on Lagrange's equations

The equations of motion for robots can be derived through the application of the Lagrange's equations of motion for nonconservative systems:

$$\frac{d}{dt}\left(\frac{\partial L}{\partial \dot{q_i}}\right) - \frac{\partial L}{\partial q_i} = \phi_i$$ (9.7)

where $L = $ Lagrangian $= T - V$

T = kinetic energy

V = potential energy

q_i = generalized joint coordinate representing the internal coordinate of the i-th joint

$\dot{q}_i$ = first time derivative of q_i

ϕ_i = generalized force applied to the i-th link; this generalized force can be thought of as the force (for sliding joints) or torque (for revolute joints) active at this joint.

Uicker (1965) and Kahn (1969) use the 4x4 rotation/translation matrices W_i introduced by Denavit and Hartenberg (see Section 9.3.2). Once the kinetic and potential energy is expressed in terms of the W_i and their derivatives, the Lagrange equation is easily applied and the generalized forces ϕ_i are found as:

$$\phi_i = \sum_{j=1}^{n} \{ \sum_{k=1}^{j} [TR \, (\frac{\partial W_j}{\partial q_i} J_j (\frac{\partial W_j^T}{\partial q_k}] \ddot{q}_k +$$

$$\sum_{k=1}^{j} \sum_{l=1}^{j} [TR \, (\frac{\partial W_j}{\partial q_i} J_j (\frac{\partial^2 W_j^T}{\partial q_k \partial q_l}] \dot{q}_k \dot{q}_l] - m_j \, \vec{g} \, \frac{\partial W_j}{\partial q_i} \vec{r}_{jo} \qquad (9.8)$$

where J_j is the 4x4 inertia matrix of the j-th link

  m_j is the mass of the j-th link

  g is the gravity vector

  $\vec{r}_{jo}$ is the distance vector between the center of mass of the j-th link and the origin
    of the reference system

  TR is the trace operator.

Unfortunately, the evaluation of Eq. (9.8) is very time consuming, because it is proportional to the fourth power of the number of links. Hollerbach (1980) has proposed a method that significantly reduces the number of operations. He first uses the Waters (1979) formulation of Eq. (9.8):

$$\phi_i = \sum_{j=1}^{n} [TR \, (\frac{\partial W_j}{\partial q_i} J_j \, \ddot{W}_j - m_j \, g^T \frac{\partial W_j}{\partial q_i} r_{jo} \qquad (9.9)$$

and proposes a forward recursion by expressing $\dfrac{\partial W_k}{\partial q_i}$ as $(\dfrac{\partial W_i}{\partial q_i}) \, W_k^i$

where $W_k^i = A_{i+1} A_{i+2} \cdots A_k$ is the matrix of transformation from the i-th into the j-th local coordinate system. By substitution, the following recursive formulation is obtained:

$$\phi_i = TR \, (\frac{\partial W_i}{\partial q_i} D_i - \vec{g} \, \frac{\partial W_i}{\partial q_i} c_i \qquad (9.10)$$

with

$$D_i = J_i \, \ddot{W}_i^T + A_{i+1} \, D_{i+1} \qquad (9.11)$$

$$c_i = \, m_i \, r_{io} + A_{i+1} \, c_{i+1} \qquad (9.12)$$

The accelerations $\ddot{W}_j$ are calculated by a backward recursion due to Waters (1979):

$$W_j = W_{j-1} A_j \qquad (9.13)$$

$$\dot{W}_j = \dot{W}_{j-1} A_j + W_{j-1} \frac{\partial A_j}{\partial q_j} \dot{q}_j \qquad (9.14)$$

$$\ddot{W}_j = \ddot{W}_{j-1} A_j + 2\dot{W}_{j-1} \frac{\partial A_j}{\partial q_j} \dot{q}_j + W_{j-1} \frac{\partial^2 A_j}{\partial q_j^2} \dot{q}_j^2 + W_{j-1} \frac{\partial A_j}{\partial q_j} \ddot{q}_j \qquad (9.15)$$

9.4.2. Methods based on Newton-Euler's equations

The Newton-Euler formulation is based on the laws governing the dynamics of rigid bodies. The procedure in this formulation is to first write the equations which define the angular and linear velocities and accelerations of each link and then write the equations which relate the forces and torques exerted on successive links while under this motion. The vector force $\overrightarrow{F}$ is given by the Newton's second law (Eq. 9.16) and the total vector torque $\overrightarrow{N}$ by Euler's equation (Eq.9.17).

$$\overrightarrow{F} = m_i \overrightarrow{\ddot{r}_i} \qquad (9.16)$$

$$\overrightarrow{N} = J_i \overrightarrow{\dot{\omega}_i} + \overrightarrow{\omega_i} \times (J_i \overrightarrow{\omega_i}) \qquad (9.17)$$

Newton-Euler's methods were first developed with respect to the fixed, inertial coordinate system (Stepanenko and Vukobratovic 1976; Vukobratovic 1978). Then methods were derived with respect to the local coordinate system. Orin et al. (1979) proposed that the forces and the torques be referred to the link's internal coordinate system. Armstrong (1979) and Luh et al. (1980) calculated the angular and linear velocities in link coordinates as well. Armstrong used coordinate systems located at the joints: Luh et al. employed coordinate systems located at the link centers of mass.

The Armstrong method is based on the hypothetical existence of a linear recursive relationship between the motion of and forces applied to one link, and the motion of and forces applied to its neighbors. A set of recursion coefficients is defined for each link and the coefficients for one link may be calculated in terms of those of one its neighbors. The accelerations are then derived from the coefficients. The method is only applicable to spherical joints, but its computational complexity is only $O(n)$, where n is the number of links. Based on this theory, Armstrong et al. (1987) designed a near real-time dynamics algorithm and implemented it in a prototype animation system To reduce the amount of computer time required, Armstrong et al. (1987) make some simplifying assumptions about the structure of the figure and can produce a near real-time dynamics algorithm. They use frames which move with the links and an inertial frame considered fixed and non-rotating. The transformation from one frame to another is done by multiplying a column vector on the left by a 3x3 orthogonal matrix. Their motion equations are as follows:

$$J_i \, \dot{\vec{\omega}}_i = - \vec{\omega}_i \times (J_i \, \vec{\omega}_i) - \dot{\vec{h}}_i + \sum_{S \in S_i} R_s \, \vec{h}_s + R_i^{I\,T} \, \vec{h}^{E}_i + m_i \vec{c}_i \times R_i^{I} \, \vec{g}$$

$$+ \, \vec{p}^{E}_i \times R_i^{I\,T} \, \vec{f}^{E}_i - m_i \vec{c}_i \times \vec{a}_i + \sum_{S \in S_i} \vec{d}_s \times R_s \, \vec{f}_s \qquad (9.18)$$

$$\vec{f}_i = -m_i \vec{\omega}_i \times (\vec{\omega}_i \times \vec{c}_i) + R_i^{I\,T} (\vec{f}^{E}_i + m_i \vec{g}) - m_i \vec{a}_i - m_i \vec{c}_i \times \dot{\vec{\omega}}_i + \sum_{S \in S_i} R_s \, \vec{f}_s \qquad (9.19)$$

The notations used are as follows:

S_i is the set of all links having link i as parent

m_j is the mass of the j-th link

$\vec{g}$ is the gravity vector

$\vec{p}^{E}_i$ is the position vector of the hinge of link i proximal to root

$\vec{f}^{E}_i$ is the external force acting on link i at $\vec{p}^{E}_i$

$\vec{h}^{E}_i$ is the external torque acting on link i

d_s is the vector from the proximal hinge of link i to the proximal hinge of its child s

$\vec{a}_i$ is the acceleration of the proximal hinge of link i

$\vec{\omega}_i$ is the angular velocity of link i

$\vec{f}_i$ is the force that link i exerts on its parent at the proximal hinge

$\vec{h}_i$ is the torque that link i exerts on its parent at the proximal hinge

R_i is a rotation matrix that converts vector representations in the frame of link i to the representation in the frame of the parent link

R_i^{I} is a rotation matrix that converts from representations in frame i to the inertial frame

J_i is the inertia matrix of the i-th link about its proximal hinge

Armstrong et al. introduce a last equation relating the acceleration at the proximal hinge of a son link s of link i to the linear and angular accelerations at the proximal hinge of i:

$$R_s \, \vec{a}_s = \vec{\omega}_i \times (\vec{\omega}_i \times \vec{d}_s) + \vec{a}_i - \vec{d}_s) \times \dot{\vec{\omega}}_i \qquad (9.20)$$

9.4.3. Methods based on Gibbs-Appel's equations

This method is based on the Gibbs formula, which describes the acceleration energy G:

$$G = \sum_{k=1}^{n} \frac{1}{2} m_i \, \vec{a}_i^2 + \frac{1}{2} \vec{\alpha}_i \, J_i \, \vec{\alpha}_i + 2(\vec{\omega}_i \times J_i \, \vec{\omega}_i) \, \vec{\alpha}_i \qquad (9.21)$$

where for link i: m_i = the link mass

$\vec{a_i}$ = the linear acceleration

$\vec{\omega_i}$ = the angular velocity

$\vec{\alpha_i}$ = the angular acceleration

J_i = the inertial tensor

Wilhelms (1987) uses the Gibbs equation to formulate the generalized force for a sliding joint k and a revolute joint for animating articulated bodies:

$$\text{sliding joint } \phi_i = \sum_{i=r}^{\text{distal}} \left(m_i a_i \left(\frac{\partial \vec{a_i}}{\partial \ddot{s_i}} \right) \right) \tag{9.22}$$

$$\text{revolute joint } \phi_i = \sum_{i=r}^{\text{distal}} \left(m_i \vec{a_i} \left(\frac{\partial \vec{a_i}}{\partial \ddot{\theta_i}} \right) \right) + \vec{\alpha_i} J_i \left(\frac{\partial \vec{\alpha_i}}{\partial \ddot{\theta_i}} \right) + \left(\frac{\partial \vec{\alpha_i}}{\partial \ddot{\theta_i}} \right)^T (\vec{\omega_i} \times J_i \vec{\omega_i}) \tag{9.23}$$

where $\ddot{\theta_i}$ is the local angular acceleration and $\ddot{s_i}$ is the local linear acceleration.

For a body with n degrees of freedom, these equations may be rearranged into a matrix formulation:

$$(\vec{q} - \vec{V}) = \ddot{\vec{c}} \tag{9.24}$$

$\vec{q}$ is the generalized force at each degree of freedom: M is an n x n inertial matrix and $\vec{V}$ a n-length vector dependent on the configuration of the masses and their velocity relative to each other. If $\vec{q}$, M and $\vec{V}$ are known, the equations can be solved for the generalized accelerations $\ddot{\vec{c}}$ (n-length vector of the local angular or linear acceleration at each degree of freedom). This corresponds to the inverse-dynamics problem.

Wilhelms (1987) used the Gibbs-Appel formulation for her animation system Deva; however this formulation does not exploit the recursive nature of the terms, and has a cost of $O(n^4)$. However, it is more general than the Armstrong approach.

9.5. Path planning and obstacle avoidance

9.5.1. Trajectory planning

Trajectory planning (Brady 1982) is the process of converting a task description into a trajectory. A trajectory is defined as a sequence of positions, velocities and accelerations. As the trajectory is executed, the tip of the end effector traces a curve in space and changes its orientation. This curve is called the path of the trajectory. The curve traced by the sequence of orientations is sometimes called the rotation curve. Mathematically the path is a vector function $\vec{p}$ (s) of a parameter s and the rotation curve may be expressed as a three scalar function $\theta_i(s)$ or a vector function $\vec{\theta}$ (s). It is also possible to define a generalized path $\vec{G}$ (s) in a six-dimensional space.

In the case of synthetic actors, motion may be described by Cartesian coordinates (end effectors) or by joint coordinates. Transformation from joint coordinates to Cartesian coordinates corresponds to the direct-kinematics problem and transformation from joint coordinates to Cartesian coordinates corresponds to the inverse-kinematics problem. Both problems and their solutions are described in Section 3.

There are several ways of defining trajectories and paths:

1. by keyframe positions
2. by explicit definition of the trajectory
3. by automatic trajectory-planning

In robotics, the geometric problem is finding a path for a moving solid among other solid obstacles. The problem is called the **Findpath** problem. One approach to the problem consists of **trying a path and testing it for potential collisions** (Pieper 1968, Lewis 1974). If collisions are detected, the information is used to select a new path to be tried. In an algorithm of this class, there are three steps (Lozano-Perez and Wesley 1979):

1. calculate the volume swept out by the moving object along the proposed path
2. determine the overlap between the swept volume and the obstacle
3. propose a new path

The first two operations are present in most geometric modellers; the third operation is much more difficult. The new path will pass on one side of the obstacle, but which side ? And if collisions are again detected, how to select a third path ? The technique becomes impracticable when the number of obstacles is large. The method often fails because each proposed path provides only local information about potential collisions. This implies local path changes when radically different paths would be better. In fact, the method works only when the obstacles are sparsely distributed so they can be dealt with one at a time.

A second class of solutions to the problem of obstacle avoidance are based on **penalty functions**. A penalty function is a function that encodes the presence of objects. The total penalty function is obtained by adding the penalties from the individual obstacles. It is also possible to refine the penalty function model by taking into account the deviation from the shortest path. In this approach, motion is planned by following local minima in the penalty function. The value of the penalty function is evaluated at each configuration by estimating the partial derivatives with respect to the configuration parameters. According to Lozano-Perez, there are severe limitations to the method, because the penalty function can only be evaluated for simple shapes like spherical robots. But the main drawback of these methods is the local information that they provide for path searching. This means that penalty functions are more suitable when only small modifications to a known path are required.

The third approach is based on the search of configurations that are free of collisions; several algorithms of this type, called **free-space** algorithms have been proposed (Widdoes 1974; Udupa 1977, Lozano-Perez 1981). Most of these algorithms are restricted to the calculation of free-space for specific manipulators like the "Stanford Arm".

9.5.2. Lozano-Perez approach

To explain the principle of free-space algorithms, we examine a 2D method introduced by Lozano-Perez and Wesley (1979). Obstacles are assumed to be polygonal, while the moving object is convex and its shape is constant. The first step consists of forming a graph. Vertices of this graph are composed of the vertices of the obstacles, the start point S and the goal point G. Arcs are defined such that a straight line can be drawn joining the vertices without overlapping any obstacle. The graph is called the visibility graph, since connected vertices can see each other. Fig. 9.9a shows an example of a visibility graph with obstacles O_i, start point S and goal point G. The shortest collision-free path from S to G is the shortest path in the graph from S to G. The method may be applied to a 2D motion in computer animation. However, it assumes that the moving object is a point; this restricts the application of the algorithm to camera motions, light motions or motions of very small objects.

Lozano-Perez and Wesley (1979) describe a way of extending the method to moving objects which are not points. In the original method, the obstacles may be considered as forbidden regions for the position of the moving point. If the moving object is not a point, obstacles may be replaced by new obstacles, which are forbidden regions for a reference point on the object.

For example, consider a circular object of radius R with the center as reference point. New obstacles are computed by expanding each boundary by R. This operation of computing a new obstacle from an original obstacle and a moving object is called **growing the obstacle by the object**. Fig. 9.9b shows the new obstacles. Now consider a rectangular object, the same procedure may be applied, as shown in Fig. 9.9c; however, some paths will be missed where the rectangular object could squeeze through by turning in the right direction, as shown in Fig. 9.9d. In fact in Fig. 9.9c no path is possible between obstacles and the goal cannot be reached. Lozano-Perez and Wesley propose some solutions to this problem, but no optimum solution is guaranteed and failure to find a solution in the graph does not necessarily mean that no safe trajectory exists.

Extensions of the method to 3D animation is possible. However, the shortest path in the graph whose node set contains only vertices of the grown obstacles is not guaranteed to be the shortest collision-free path. Lozano-Perez and Wesley propose to add vertices on obstacle edges. It should also be noted that the 3D growing operation is generally time-consuming.

Lozano-Perez (1981) formally describes a method for computing the exact configuration of space obstacles for a cartesian manipulator under translation. He introduces a representation called **Cspace**; the Cspace is the six-dimensional space of configurations. A configuration is a single six-dimensional vector specifying the position and orientation of the solid.

Brooks (1983) suggests another method called the **freeway** method. His algorithm finds obstacles that face each other and generates a freeway to passing between them. This path segment is a generalized cylinder. A freeway is an elongated piece of free-space that describes a path between obstacles. This freeway may be described as overlapping generalized cones; it is essentially composed of straight lines with left and right free-space width functions, which could easily be inverted. A generalized cone is obtained by sweeping a two-dimensional cross section along a curve in space, called a spine, and deforming it according to a sweeping rule.

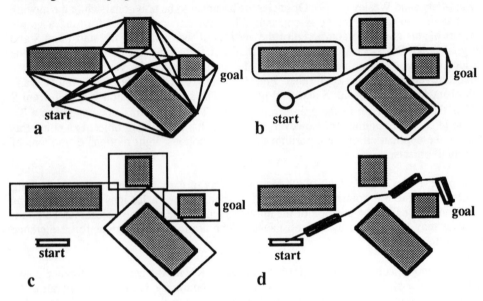

Figure 9.9. a. Obstacles, start point, goal and path; **b.** Circular object, grown obstacles and path; **c.** Rectangular object and grown obstacles; no path is possible between obstacles; **d.** Rectangular object; a path is possible when rotating the object

9.6. Behavioral animation

With the development of the simulation of living creatures, researchers have introduced rather recently into computer animation the concept of automatic motion control using mechanics-based techniques such as dynamic simulations and robotics-based task planning. However, motion of 3D characters is not simply a matter of mechanics: you cannot walk exactly the same way from the same bar to home twice. Behavioral animation corresponds to modeling the behavior of characters, from path planning to complex emotional interactions between characters. In an ideal implementation of a behavioral animation, it is almost impossible to exactly play the same scene twice.

Reynolds (1987) introduces a distributed behavioral model to simulate flocks of birds, herds of land animals, and schools of fish. The simulated flock is an elaboration of a particle system with the simulated birds being the particles. A flock is assumed to be the result of the interaction between the behaviors of individual birds. Working independently, the birds try both to stick together and avoid collisions with one another and with other objects in their environment. In a module of behavioral animation, positions, velocities and orientations of the actors are known from the system at any time. The animator may control several global parameters: e.g. weight of the obstacle avoidance component, weight of the convergence to the goal, weight of the centering of the group, weight of the velocity equality, maximum velocity, maximum acceleration, minimum distance between actors. The animator provides data about the leader trajectory and the behavior of other birds relative to the leader. A computer-generated film has been produced by Symbolics using this distributed behavioral model: *Breaking the ice.*

For example, in the task of walking, everybody walks more or less the same way, following more or less the same laws. This is the "more or less" which will be difficult to model. And also a person does not walk always the same way everyday. If the person is tired, or happy, or just got some good news, the way of walking will appear slightly different. So in the future, another big challenge is open for the computer animation field: to model human behavior taking into account social differences and individualities.

Renault et al. (1990) propose a prototype to give to the synthetic actor a vision of his environment. It seems to be a simple and stable solution for providing him the information he needs to react and behave himself. The big problem is to transform this information into knowledge. The prototype has shown that this vision-based approach to behavioural animation is a workable solution. An advantage of the synthetic vision is that we may learn from research on human vision. The synthetic environment chosen for these trials is a corridor containing several obstacles of various sizes. The synthetic actor may be placed at any location of the corridor and with any look direction; he will move along the corridor avoiding the obstacles. Path planning based on synthetic vision may be dynamically performed without using powerful mathematical algorithms. The system is able to avoid collisions with movable objects, what is not possible with well-known robotics algorithms of path-planning as described in Section 9.5.

Acknowledgment

This work has been partly supported by the Swiss National Science Foundation.

References

Armstrong WW (1979) Recursive Solution to the Equations of Motion of an N-Link Manipulator, *Proc. 5th World Congress Theory Mach.Mechanisms*, Vol.2, pp.1343-1346

Armstrong WW, Green M, Lake R (1987) Near real-time Control of Human Figure Models, *IEEE Computer Graphics and Applications*, Vol.7, No 6, pp.28-38

Boulic R, Magnenat-Thalmann N, Thalmann D (1990) Human Free-Walking Model for a Real-time Interactive Design of Gaits, *Computer Animation '90*, Springer-Verlag, Tokyo

Brady M (1982) Trajectory Planning, in: *Robot Motion: Planning and Control*, MIT Press, Cambridge, Mass.

Brooks RA (1983) Planning Collision-Free Motions for Pick-and-Place Operations, *International Journal of Robotics*, Vol.2, No4, pp.19-26

Cohen MF (1989) Gracefulness and Style in Motion Control, *Proc. Mechanics, Control and Animation of Articulated Figures*, MIT

Denavit J, Hartenberg RS (1955) A Kinematic Notation for Lower Pair Mechanisms Based on Matrices, *J.Appl.Mech.*, Vol.22, pp.215-221.

Featherstone R (1983) Position and Velocity Transformations Between Robot End-Effector Coordinates and Joint Angles, *Intern. Journal of Robotics Research*, Vol.2, No2, pp.35-45.

Gurfinkel VS, Fomin SV (1974) Biomechanical Principles of Constructing Artificial walking Systems, in: *Theory and Practice of Robots and Manipulators*, Vol.1, Springer-Verlag, NY, pp.133-141

Hemami H, Farnsworth RL (1977) Postural and Gait Stability of a Planar Five Link Biped by Simulation, *IEEE Trans. on Automatic Control*, AC-22, No3, pp.452-458.

Hewitt CE (1977) Viewing Control Structures as Patterns of Passing Messages, *Journal of Artificial Intelligence*, Vol.8, No3, pp.323-364

Hollerbach JM (1980) A Recursive Lagrangian Formulation of Manipulator Dynamics and a Comparative Study of Dynamics Formulation Complexity, *IEEE Trans. on Systems, Man, and Cybernetics*, Vol. SMC-10, No11, pp.730-736

Hollerbach JM, Sahar G (1983) Wrist-Partitioned, Inverse Kinematic Accelerations and Manipulator Dynamics, *Intern. Journal of Robotics Research*, Vol.2, No4, pp.61-76.

Kahn ME (1969) *The Near-Minimum-Time Control of Open-Loop Articulated Kinematic Chains*, Stanford Artificial Intelligence project, AIM-106

Korein J, Badler NI (1982) Techniques for Generating the Goal-directed Motion of Articulated Structures, *IEEE Computer Graphics and Applications*, Vol.2, No9, pp.71-81

Korein JU (1985) *A Geometric Investigation of Reach*, MIT Press

Lewis (1974) *Autonomous Manipulation on a Robot: Summary of Manipulator Software Functions*, Technical Memo 33-679, Jet Propulsion Laboratory

Lozano-Perez (1982) Task Planning in: Brady M (Ed.) *Robot Motion: Planning and Control*, MIT Press, Cambridge, Mass.

Lozano-Perez T (1981) Automatic Planning of Manipulator Transfer Movements, *IEEE Trans. on Systems, Man and Cyberbetics*, Vol. SMC-11,No10, pp.681-698.

Lozano-Perez T, Wesley MA (1979) An Algorithm for Planning Collision-Free Paths Among Polyhedral Obstacles, *Comm.ACM*, Vol.22, No10, pp. 560-570.

Luh JYS, Walker MW, Paul RPC (1980) On-line Computational Scheme for Mechanical Manipulators, *Journal of Dynamic Systems, Measurement and Control*, Vol.102, pp.103-110.

Magnenat-Thalmann N, Thalmann D (1983) The Use of High Level Graphical Types in the MIRA Animation System, *IEEE Computer Graphics and Applications*, Vol. 3, No 9, pp. 9-16.

Magnenat-Thalmann N Thalmann D (1987) The Direction of Synthetic Actors in the film Rendez-vous à Montréal, *IEEE Computer Graphics and Applications*, Vol. 7, No 12, pp.9-19.

Magnenat-Thalmann N, Thalmann D, Fortin M (1985) MIRANIM: an extensible director-oriented system for the animation of realistic images, *IEEE Computer Graphics and Applications*, Vol.5, No3, pp. 61-73

McMahon T (1984) Mechanics of Locomotion, *Intern. Journ. of Robotics Research*, Vol.3, No2, pp.4-28.

Miura H and Shimoyama I (1984) Dynamic Walk of a Biped, *International Journal of Robotics*, Vol.3, No2, pp.60-74.

Orin D, McGhee R, Vukobratovic M, Hartoch G (1979) Kinematic and Kinetic Analysis of Open-Chain Linkages Utilizing Newton-Euler methods, *Mathematical Biosciences*, Vol.31, pp.107-130.

Pieper (1968) *The Kinematics of Manipulators under Computer Control*, Ph.D Dissertation, Stanford University, Computer Science Department

Popplestone RJ, Ambler AP and Bellos IM (1980) An Interpreter for a Language for Describing Assemblies, *Artificial Intelligence*, Vol.14, pp.79-107

Raibert MH, Sutherland IE (1983) Machines that Walk, *Scientific American*, Vol.248, No1, pp.44-53

Renault O, Magnenat-Thalmann N, Thalmann D (1990) A Vision-Based Approach to Behavioural Animation, *Visualization and Computer Animation Journal*, John Wiley, Vol,1, No1 (July 1990)

Reynolds CW (1982) Computer Animation with Scripts and Actors, *Proc. SIGGRAPH'82, Computer Graphics*, Vol.16, No3, pp.289-296.

Reynolds C (1987) Flocks, Herds, and Schools: A Distributed Behavioral Model, *Proc.SIGGRAPH '87, Computer Graphics*, Vol.21, No4, pp.25-34

Salisbury JK, Craig JJ (1982) Articulated Hands: Force Control and Kinematic Issues, *Intern. Journal of Robotics Research*, Vol.1, No1, pp.4-17.

Saunders JB, Inman VT, Eberhart (1953) The Major Determinants in Normal and Pathological Gait, *Journal of Bone Joint Surgery*, Vol.35A, pp.543-558

Stepanenko Y and Vukobratovic M (1976) Dynamics of Articulated Open Chain Active Mechanisms, *Mathematical Biosciences*, Vol.28, pp.137-170.

Turner R, Gobbetti E, Balaguer F, Mangili A, Thalmann D, Magnenat Thalmann N (1990) An Object-Oriented Methodology using Dynamic Variables for Animation and Scientific Visualization, *Proc. Comp. Graphics Intern. '90*, Singapore, Springer-Verlag, Tokyo

Udupa SM (1977) Collision Detection and Avoidance in Computer Controlled Manipulators, *Proc. IJCAI-5*, MIT, pp.737-748

Uicker JJ (1965) *On the Dynamic Analysis of Spatial Linkages Using 4x4 Matrices*, Ph.D Dissertation, Northwestern University, Evanston, Illinois

Vukobratovic M (1978) Computer method for Dynamic Model Construction of Active Articulated Mechanisms using Kinetostatic Approach, *Journal of Mechanism and Machine Theory*, Vol.13, pp.19-39

Waters RC (1979) *Mechanical Arm Control*, A.I. Memo 549, MIT Artificial Intelligence Laboratory

Widdoes LC (1974) *Obstacle Avoidance, a Heuristic Collision Avoider for the Stanford Robot Arm*, Technical Memo, Stanford University Artificial Intelligence Laboratory

Wilhelms J (1987b) Using Dynamic Analysis for Realistic Animation of Articulated Bodies, *IEEE Computer Graphics and Applications*, Vol.7, No 6, pp.12-27

Zeltzer D (1982) Motor Control Techniques for Figure Animation, *IEEE Computer Graphics and Applications*, Vol.2, No9, pp.53-59.

Zeltzer D (1985) Towards an Integrated View of 3D Computer Animation, *The Visual Computer*, Vol.1, No4, pp.249-259.

10

Graphic Representation of Numerical Simulations

Silvio Merazzi
SMR Engineering & Development

Ernesto Bonomi
Ecole Polytechnique Fédérale de Lausanne

10.1 Introduction

Information processing in applied science and engineering usually results in huge amounts of data. These "numbers" are basically produced by two distinct sources, namely experimental data acquisition systems and numerical simulations. All these data have one thing in common: they must be reduced, augmented or transformed such that meaningful answers emerge.

Within the context of this presentation, we have concentrated our efforts on the processing of data resulting from the numerical solution of large 3-dimensional problems, i.e. problems with hundreds of thousands of computational nodes. A computer simulation of a physical process traditionally starts with a mathematical model consisting of differential equations that govern macroscopical quantities varying in space and in time. The discretization by means of finite elements, finite volumes or finite difference methods leads to large sets of algebraic equations. Typically, in fluid mechanics, the current state at each node is described by the three components of the velocity vector, the density and the pressure. Another example is the stress analysis in solids where one deals with node-wise defined displacement fields and element-wise defined stress tensor fields.

The need for realism during the graphical rendering of simulations implies improved display techniques in order to visualize and to efficiently analyze physical processes that occur in complex geometries. Graphical tools that simply accept arrays of numbers to construct multi-dimensional curves are no longer sufficient. The complexity of large 3-dimensional models demands dedicated criteria to retrieve data from a somewhat structured data base for depicting the spatial and time behaviour of significant scalar and vector fields. Examples of such criteria are selection by elements, by nodes, by slices (plane cuts), by iso-lines or by iso-surfaces.

Modern engineering workstations, specially those equipped with sophisticated graphical processors, offer possibilities engineers could only dream of some years ago. New hardware and software environments require some quite radical changes in software development. As an example, take 3-dimensional graphics display techniques: some years ago we had to write large programs to produce "hidden line" plots or shaded contour plots, all of them based on "line-oriented" decomposition methods. Modern graphic workstations virtually eliminate these problems, since such basic operations are done in the computer's hardware. A couple of system calls produce the desired effect. However, these new facilities require working with polygones rather than with lines, a concept which implies a radical change in the approach.

In what follows we present some current developments of the ideas outlined above. Section 2 gives an introduction to the data definitions of the B2000 system as desribed by Stehlin and Merazzi (1985) and of the ASTRID system (Merazzi, 1988). Section 3 explains the ASTRID and B2000 user interface. Section 4 discusses the various visualization techniques. Finally, section 5 explains some aspects of the implementation of the graphic visualization techniques.

10.2 Data Structures and Data Management

The design of integrated finite element systems relies on common rules for the description of the data structures and for the data flow during the various stages of a simulation process. Data structure and data definition must be designed for the the numerical solution of partial differential equations in a multi-machine environment. Typical data structures are described by Stehlin and Merazzi (1985) and by Merazzi (1988). Figure 10.1 schematically depicts a client-server-based multi-machine environment.

A finite element grid consists of an assembly of polyhedra, usually thetahedral or hexahedral, modelling the computational spatial domain. The general description of the geometrical properties of elements, i.e. type, list of node indices, list of faces etc., suggests separate lists for each singular element, a concept which causes an unstructured numbering of the elements. The absence of structure in the grid implies a consequent design of the data structure and the maintaining of element and node pointers. Figure 10.2 (see color section) displays an unstructured mesh.

In another approach, the tesselation of the spatial domain is obtained by assembling a collection of structured subdomains. A structured sudomain is defined as a grid which is

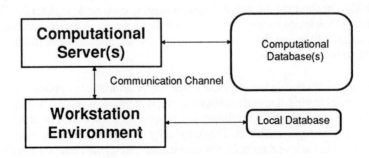

Figure 10.1 - Multi-Machine computing environment

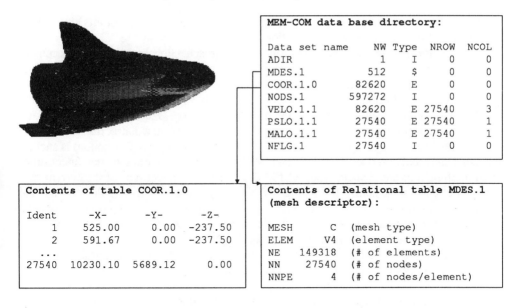

Figure 10.4 - Mesh and corresponding MEM-COM data structure

homeomorphic to a regular lattice, see Bonomi (1989). Obviously, this approach is less general than the unstructured approach. However, the element grid of each subdomain - usually hexahedral cells - inherits the structured numbering of its lattice of reference, allowing for an automatic description of the neighbourhood of each element. Moreover, it guarantees band-structured matrices which are the prerequisites for efficient processing in parallel vector computers. Unfortunately, the automatic decomposition of an arbitrary spatial domain into purely structured subdomains still is an unsolved problem. An eloquent example, suggested by Droux (see Desbiolles et. al., 1989), is the decomposition of the patterns obtained by intersecting two cylinders as depicted in Figure 10.3 (see color section).

   An intermediate grid type consists of unstructured meshes made out of elements having identical geometrical and physical attributes. Compared to pure unstructured meshes this description is simplified. Figure 10.2 (see color section) displays such an intermediate mesh type made out of facets that originate from thetahedral elements.

   The data structure must in any case support all types of meshes, i.e. unstructured meshes, structured meshes and semi-structured meshes, since structured meshes will generate unstructured 2-dimensional meshes when sliced.

   The data design and the data flow management of the ASTRID/B2000 systems are layed out for common memory parallel computers, see Gruber et. al. (1989). The self-descriptive data representation on the data bases facilitates network management . The problem of data transfer is then reduced to the actual communication and is solved by special system calls.

   The MEM-COM Data Manager (SMR, 1988) and the Dynamic Memory Manager (Merazzi, 1988) provide for very efficient memory and file management of large vectors, matrices and, to some extent, relational tables in a FORTRAN77 or C programming environment. MEM-COM manipulates data objects symbolically, i.e. by name and attributes. MEM-COM does not support structures (as defined in the C language) directly since constructs of this kind violate the rule of self-descriptive data representation.

The data classification is organized in a top-down approach. Global relational tables contain "tables of content" which lead the way through the data. Figure 10.4 displays a summary of some MEM-COM data sets that point to the actual data. Set **ADIR** (Analysis DIRectives) is a relational table that informs on the contents of the data file such as the type of problem, the number of subdomains, the total size of the problem and so on. Data set **COOR.1.0** (i.e. the COORdinates of subdomain 1 and computational cycle 0) points to a matrix that contains the node coordinates of the computational mesh. Set **MDES.1** (i.e. Mesh DEScription) is another relational table that describes the subdomain. MESH contains the mesh type flag (an unstructured mesh of constant element type). ELEM describes the element type of the current subdomain (tetrahedral elements). Entries NE and NN designate the number of elements and the number of computational nodes. The **NODS** data set contains the element connectivity table which, for constant mesh types, has a fixed column size. Note that for unstructured meshes the column size is variable and that for structured meshes NODS is not needed.

More complicated constructs, such as parametrized lines, surfaces of volumes, etc., can be described similarly by combining MEM-COM data structuring techniques.

Data is brought in and out of memory by means of a dynamic memory management system. As an example, the node coordinate table COOR.1 are copied in memory:

```
address = dmmget('COOR.1.0')
```

The procedure 'dmmget' loads the table in dynamic memory and returns an address pointer. Relational table entries are retrieved in a similar way:

```
NN = gettable('NN')
```

The procedure 'gettable' returns the integer constant 'NN' from the table currently loaded in memory.

All examples listed above pertain to coded sections of the program modules. Similar calls can be made interactively from the command interpreter (see next section).

Since the generation of a model and the analysis of the results and, to some extent, the actual control of the solution process, are performed on a workstation, some additional features are imposed on the design. The model must be described with a minimum amount of information. It is more efficient to recompute data than to store it. The definition of similar data bases should be possible on all computers in the network. Typically, the initial data base generated on the workstation is re-used for processing results that arrive from the computational server. Only new data or data that have been modified are transmitted. All computers involved in the problem solution have access to the ASTRID system. When needed for the examination of results or for remeshing the model (Bonomi, 1989), gradients may be recomputed locally on the workstation for example, rather than transmitted from the vector computer. Thus, the amount of information exchanged between the workstation and the vector computer must be minimized.

10.3 User Interfacing

This section gives a short account of some techniques of communication with a program, and it resumes the methods selected for the ASTRID/B2000 system.

The most elementary way of conversing with an interactive program is by means of a "question-and-answer" game. The program asks for some input. Based upon the answer, another set of of questions appears etc. More sophisticated systems address portions of the video screen. These "screen masks" give a better overview than pure line-by-line input. The concept of "question-and-answers" can now be extended to graphics. Instead of pure text, clusters of text boxes and symbols ("icons") now appear on the terminal. Of course, graphical systems do not work well on traditional video terminals, because they lack sophistication and transmission speed. Graphical systems (user interfaces) are practically only feasible on the high-resolution, bit-mapped graphic screens offered by modern workstations. These devices also provide for "windows" through which one converses with the computer. Each window is assigned to a separate process, such as editing, graphics, computing etc., thus separating activities more clearly.

Window systems have become a generally accepted standard in engineering workstations. APOLLO's Display Manager or the X-Windows system are examples of such systems. Graphical user interfaces have gained wide popularity because they facilitate the entry to many application programs.

There is no doubt about the usefulness of icons, scrollbar menues etc. But they have their limits. What we require is a command language similar to the UNIX shell or to a programming language like C or LISP. Consequently, we need both a command-language driven user interface and, on top of it, an optional icon-driven user interface.

10.3.1 A Rewiew of Some User Interface Tools

APOLLO Computer's DOMAIN/Dialog system and SUN's SUNView are integrated proprietary systems. They provide for user interface "toolboxes" but they are not portable. Dialog allows for the complete separation of the application program from the user interface. The application program and the user interface can thus be developed almost separately, but the data exchange between the user interface and the application is not solved by Dialog, since data exchange is language-dependent.

X-Windows: to write user interfaces implies large investments in code development, even if tools like Dialog are available. The need for standardization has led to the development of the X-Windows standard. However, the X-Windows is not a user interface management system.

Another standard adopted by some vendors is the NeWS system. Unlike X-Windows, NeWS is based on an extension of an established graphic language, the PostScript language. NeWS also requires a PostScript interpreter. Both X-Windows and NeWS are only accessible via the C language. Like X-Windows, NeWS is not a user interface management system.

New standards, like the OSF MOTIF system, will eventually become available.

10.3.2 The ASTRID/B2000 Command Language

The ASTRID/B2000 user interface is somewhat different from the systems outlined above. One of the major particularities is that the ASTRID system must be programmable interactively. In this sense it is simliar to a UNIX shell. Any graphical user interface can be placed on top of the interpreter and it will generate keywords or scripts from the graphical symbols selected.

If the user interface is well designed, you can let it do most of the job of input definition. For example, the ASTRID geometry generation module accepts very few commands, among these "define a point" or "define a line" i.e. a sequence of points. To produce a sequence of points which lie on a circle we may either program some piece of code within the module that accepts the commands to generate these points, or we may write a procedure within the command language that generates points and transmits them to the application. A command script to generate a circle is given below:

```
(r=10.)
( x(ang)=r*cos(0.01745*ang) )
( y(ang)=r*sin(0.01745*ang) )
(delta=5.) (ip=0)
while  (ang<90.) ((ip=ip+1)
           CP (ip) (x(ang)) (y(ang)) 0.0;
           (ang=ang+delta) )
```

Instructions placed between parentheses define or evaluate expressions. The bolded line above contains the actual instructions to be transmitted to the application (CP, i.e. Create Point). We therefore tend to put more power and flexibility into the interface instead of blowing up the application programs.

The user interface must also be able to give, at any time, direct access to the data base.

Figure 10.5 displays the interaction of the user-interface with an application. The application dictates all data aquisition tasks and data, but it does neither make any assumptions on how operating instructions and data should be transmitted, nor does it specify where these instructions and data are stored. The application simply receives instructions and data according to the application dictionary. Messages are sent back to the user interface if necessary. All data generated by the user are stored in the variable stack or on the data base. The application retrieves data from the variable stack or from the data base only when needed. User interface tasks are activated when the appropriate events have been detected (keyboard input, mouse activated, application sends message etc.).

The user interface is primarily designed for operation in command language mode. As soon as an application becomes stable, i.e. most of the application dictionary has been defined, the menu part of the user interface can be written. The reasons for proceeding in steps are two-fold. Firstly, it is much easier to debug an application in command mode. Secondly, the programming effort for getting nice icons, menues etc. is quite considerable. It is more efficient to make this effort once all tasks have been defined.

Some other important reasons for relying on a command language are repetitive operations, such as the production of a series of similar diagrams, or the generation of animated sequences. Although one can imagine providing some pre-programmed animations, these would always lack the necessary flexibility. Animated sequences (or actual pseudo-animations, since

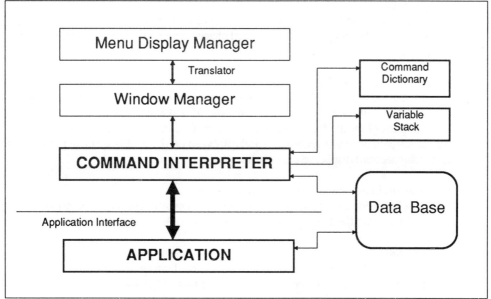

Figure 10.5 - User interface and application setup

the dynamic process does not operate in real-time) can be obtained very elegantly by means of command scripts.

The following example generates temperature contour plots on the surface of a 3-dimensional object for a sequence of solutions at given integration times:

```
(icmax=120)              !! max. computational cycle
(ic=1)                   !! current computational cycle
while   (ic<icmax)
        ( selfield TEMP cycle=(ic); plot;
        (ic=ic+1) )
```

The overhead due to the interpretation of the commands within an animation script does not penalize the execution speed too heavily: on a Silicon Graphics 4D70GT we still get a throughput of around 15000 to 30000 polygones per second, depending on the polygon size (in pixels).

10.4 Visualization of Data

As we primarily deal with large 3-dimensional objects (computational grids) the graphic modules must be capable of rendering many polygones and lines very efficiently. The BASPL visualization tool (Merazzi, 1988) has been devised to operate on graphical workstations with substantial hardware support, i.e. transformations, hidden-face elimination and light-models are provided by the hardware.

Graphical visualization techniques of finite-element simulations include standard techniques such as:

- **Shape plots**
  They include wireframe plots and solid model plots. Figures 10.2 and 10.3 (color section) display solid model plots. Deformed shapes may be displayed superimposed to the shape.

- **Color contour plots of scalar fields**
  The model, parts of it, or slices are shaded according to the intensity of the scalar field. Figure 10.6 (color section) gives an example of a shaded contour plot of both the selected parts of the model and a plane intersection (slice). Contour lines are computed for all selected elements (polygons) and displayed on the visible surface. Figure 10.7 (see color section) displays a contour line plot.

- **Contour surfaces (iso-surfaces) of scalar fields**
  Scalar fields inside solid models can be represented by surfaces that correspond to constant values of the field, see Figure 10.8 (color section).

- **Vector plots**
  Velocity fields, force fields or displacement fields, both on the selected parts of the model and on slices, can be visualized by vector plots. Optional color attributes of the vectors may be assigned to the amplitudes, time etc.

- **Particle tracks**
  Particle tracks result from the integration of a velocity field or from the integration of the equation of motion of a point mass placed in a force field. Tracks may be displayed statically as solid lines or as animated sequences. Scalar values may be displayed in continuous color tones along the path. Figure 10.9 (color section) displays tracks of different particles moving in a velocity field. The local pressure field which the particles cross, is symbolized by continuous colors.

In the following paragraphs we give an account of a typical post-processing session. The geometrical model and some results of compressible Euler flow computations around a body placed in a flow field are available on a data base file. The computational domain consists of some 150000 tetrahedral elements assembled in an unstuctured (constant) mesh. The computations were performed with a code developed at Avions Marcel Dassault/Bréguet Aviation.

The first part of the session consists of visualizing the model. A specific selection criterion must be applied in order to extract significant parts of the geometry, i.e. the facets of the elements which are assigned to the surface of the body in the flow field must be extracted. In our case we scan all facets of all elements retaining only those facets whose vertices correspond to a "node flag", i.e. a code assigned to each node of the domain. In the BASPL command language these requests look like

```
selelem
   subdomain=1   allelements   for   'nodeflag=1'
end
```

The facets selected by this command are displayed in Figure 10.2 (color section). The results from the flow computations include 3-dimensional velocity fields available at the nodes of the computational mesh. It is thus possible to compute particle tracks by interpolating the velocity field and by integrating the equation $\underline{v} = d\underline{s}/dt$, given an initial position $\underline{s}_0$. The integration is rather time-consuming. It may therefore be performed either separated from the graphics program, within the graphics program or on the computational server hooked up to the graphics program as schematically displayed in Figure 10.1. Section 5 gives an account of the particle integration method. Once computed, the particle tracks are stored on the common data base. The raw tracks can then be displayed or further processed. To obtain more realistic performance animated sequences require the synchronisation of all particles for all paths, i.e. a common time step must be determined. Figure 10.9 (color section) displays the particle tracks in the form of statically displayed solid lines. Other display methods consist of viewing the particles as they move along the path in animated sequences.

10.5 Some Implementation Notes

BASPL has been designed for graphics workstations that support 3-dimensional graphics by their hardware. We therefore have decided to separate the graphics modules completely from any window manager, i.e. we are not calling X-Windows graphics primitives or PostScript functions, since they are far too slow. The window manager simply defines a rectangular area on the screen directly tied to applications which make use of 3-dimensional graphics libraries such as the Silicon Graphics GL library or PHIGS.

BASPL works almost exclusively with 3-dimensional polygon display lists. Selected elements (cells) are broken down in polygones which are copied to the display list. The display list itself is a linked list in dynamic memory. Each entry in the list is a template of the form

```
next_polygon_pointer
n_of_points_of_polygon
element_origin_id
subdomain_origin_id
connectivity_list[1..n_of_points_of_polygon]
( coordinates[1..3:1..n_of_points_of_polygon] )
polygon_normal[1..3]
```

The linked list is necessary since we do not know a priori how large the polygon list is going to be. The polygon list itself defines the geometry of the model. To visualize solution fields the corresponding tables in the data base must be copied to dynamic memory. The connectivity list in the above template then points to the fields.

Pure element-based definitions of polygon display lists are often not sufficient to produce the desired visual effects. If, for instance, the outer shell of a body is to be visualized as a semi-transparent membrane, then all facets of all elements inside the shell must be eliminated. A fast polygon and line sorting method has been devised and implemented in BASPL by Bonomi and Merazzi, see Merazzi (1988).

Another selection of display lists is based on plane intersections of 3-dimensional meshes. The intersection patterns then define new display lists on which scalar and vector fields may be interpolated. Figure 10.7 (color section) gives such an example with both an element-based and an intersection-based display list selection.

The computation of iso-surfaces is basically identical to the determination of a plane intersection polygon list. Instead of giving the equation of a plane that defines an intersection pattern (polygon) for each element, the scalar field values at the vertices of an element are prescribed.

The determination of tracks of particles moving in a velocity field requires an appropriate data description, especially if the tracks are computed in unstructured meshes, such as the one displayed in Figure 10.4 (color section). The problem is divided into 3 sub-problems: (1) For a given initial position of the particle find the start element for the integration; (2) Integrate the interpolated velocity field within the element until the particle hits one of the faces spanning the element; (3) Find the element that is connected to that face and restart the integration. Bonomi and Merazzi (1989) give an account of the algorithm implemented for unstructured grids.

The definition of color palettes, especially during continuous shading visualization processes, is not evident. Models based on the variation of the hue, i.e. the spectrum of visible light, may not always give satisfactory results. If the spectrum is divided linearly in many segments (typically 1000) some color ranges, in particular green tones, may not be distinct enough. Nonlinear laws must then be applied.

Another problem related to continuous shading may appear if hardware shading techniques are involved: if the field values at the vertices, i.e. the shading values, of a polygon exhibit important variations of the current palette, the interpolated colors inside the polygon will be wrong since too few interpolation vertices are available.

10.6 Concluding Remarks

The present paper describes an attempt to integrate data structuring techniques, analysis methods and visualization systems. The approach offers a solution to engineers and physicists to infer useful information from large and costly computer simulations. The approach is one of many possible ones within the context of today's computer systems and programming tools. Common data structures in a multi-machine environment and data flow management designed for numerical simulations are some key elements of the systems presented.

Acknowledgements

Michel Flück has developed the expression evaluation module of the ASTRID system. Tam Cong Tran has given useful hints during the development of the graphical parts of the programs discussed herein. Both work for the EPFL, Lausanne. The HERMES space shuttle prototype finite element modelv data displayed in Figures 10.2, 10.6, 10.8 and 10.9 appear courtesy of Avions Marcel Dassault/Bréguet Aviation.

References

Bonomi, E. (1989). Adaptive Meshing of Surfaces by Structured Grids using Molecular Dynamics. Proc. of the 5th international symposium on numerical methods in engineering, Lausanne, September 1989. Comp. Mech. Publications, Springer Verlag, 1989.

Bonomi, E. and Merazzi S. (1989). Particle Tracing in Unstructured Flow Field Meshes. GASOV Report 18, EPFL, Switzerland.

Desbiolles J.L., Droux J.-J., Hoadley A.F.A., Rappaz J. and Rappaz M. (1988). Modelling of Solidification Processes. Proc. of the 5th international symposium on numerical methods in engineering, Lausanne, September 1989. Comp. Mech. Publications, Springer Verlag, 1989.

Gruber, R. et. al. (1989). Structured Finite Element and Finite Volume Programs Adapted to Parallel Vector Computers. To appear in Computer Physics Reports.

Merazzi, S. (1988). The Astrid Data Structure and Processor Handbook. GASOV Report 12, EPFL, Switzerland.

Merazzi, S. (1988). The BASPL2000 Postprocessing System. GASOV report 14, EPFL, Switzerland (1988).

SMR Corporation (1988). The MEM-COM User Manual. SMR Corporation, Bienne, Switzerland.

Stehlin, P. and Merazzi, S. (1985). B2000 Data Access Methods and Data Base Description Handbook. SMR Corporation, Bienne, Switzerland.

11

Visualization of Flow Simulations

R. Richet, J.B. Vos, A. Bottaro and S. Gavrilakis
Swiss Federal Institute of Technology, Lausanne

The use of computer graphics to visualize the results from fluid flow simulations has become an integral part of fluid dynamics research. With reference to three specific areas - hypersonic flows, flow in hydraulic turbines and turbulence - the use of graphics is described and an example of a recent software package is briefly introduced.

11.1. Introduction

Until about thirty years ago fluid dynamics research was essentially based on two methods: experimental and theoretical, the latter often referred to as analytical. In the early days, the application of the knowledge of fluid dynamics in engineering, as for example on aircraft design, was dependent on correlations which had been established by experiments. With the continuous proliferation of powerful computers a third approach, called either computational or numerical, has become widely available, and at present numerical work is complementing experimental and analytical efforts. Although numerical methods can never substitute experiments, one may observe a shift towards greater reliance on numerical simulations for design and research purposes, using experimental techniques to verify these designs and/or to validate numerical results.

Visualization of simulation data is an important aid to fluid dynamics research. Until the early 1980's two-dimensional (2D) flow simulations were most common, and did not require highly specialized graphical tools to analyze the results. With the present generation of supercomputers, like the Cray 2, Cray Y-MP and NEC SX2, three-dimensional (3D) flow simulations have become a common practice. This has changed the requirements for graphical tools due to the increase in volume and complexity of the data generated by these simulations. Today highly specialized graphical workstations are available and are used to put these data in a form which can be more easily assimilated by the human mind. The term visualization is now used to denote the broad area of research which aims at maximizing the effectiveness of the pattern recognition abilities of the human visual system. The main sources of difficulty in visualization work of flow simulations are associated with the spatial complexity and the dimensionality of the data. Various techniques of data reduction and feature enhancement are in use, but they tend to be application specific.

A 3D flow simulation can be divided into three distinct phases. To start with one has to generate the grid for the chosen computational domain. The construction of this grid is important since an "unsuitable"[1] choice can degrade the accuracy of the final result. During this phase the analyst should have some insight of what the correct solution looks like. Graphical workstations are often used to create and visualize the grid, but the choice of an optimal grid rests on human judgement. In the second phase, the discretized equations of fluid motion are solved. Very often a marching procedure in time or space is used to solve these equations, and visualization of a selection of intermediate results would be helpful to check the progress of the computation. However, this has not been possible yet due to the large amount of data that needs to be transferred over the network connecting the supercomputer and the graphical workstation. Moreover, it is only recently that graphical workstations are able to process large amount of data in a short time. The final phase in a fluid dynamics simulation is the postprocessing, i.e., the displaying and interpretation of the computed results, and it is here that graphical workstations have become a common tool. Interesting examples of such visualizations are reported, among others, in the articles by Buning and his co-workers given in the references.

The output of flow simulations is usually stored in terms of the values of the velocity vector components, the scalar pressure field, and - if applicable - temperature and/or density fields. For transient flow simulations it is usually necessary to store a time history for each variable as well. The method of graphical representation of such data depends strongly on the nature of the problem being studied. If the flow field has single dominant features, like a shock wave or vortex, visualization can be done using existing software tools. However in the case of flow fields in complex geometries the data to be displayed is usually reduced according to some selection criteria that may be intuitive or may have been suggested by experimental or theoretical studies.

In the following sections an account is given on the use and role of graphics in computational fluid mechanics, and an outline of the main features of the graphics package developed and in use at IMHEF is described. Three selected physical applications are presented, and the use and requirements of graphics in the simulation of these problems is discussed.

11.2. Computational Fluid Dynamics

Computational Fluid Dynamics (CFD) comprises all fields of research and engineering dealing with the numerical simulation of fluid flows. It is used as a tool to increase the physical understanding of complex flow phenomena (i.e. turbulent flows, chemically reacting flows, etc.) or as one of the design instruments in engineering applications (i.e. design of airplanes, chemical reactors, cooling systems for power stations, etc.). CFD is used to simulate a wide variety of flows, as for example environmental flows, reactive flows in combustion engines, but also magneto-hydrodynamic flows which occur between galaxies in space.

---

[1]   For unsuitable it is intended a mesh which does not possess certain qualities of regularity, quasi-orthogonality, etc. The reader is referred to Thompson (1982).

11.2.1 Role of graphics in CFD

As mentioned above the first step of a numerical flow simulation consists in defining an appropriate discretization grid system. To perform this task new CAD/CAM methods are proving very useful for grid generation within complex geometries. For numerical accuracy it is necessary that grid spacing is equidistant or almost equidistant, and the grid lines orthogonal or quasi-orthogonal to each other. For practical 3D geometries it is difficult to devise automated grid generation algorithms that converge to an optimum solution, and therefore interactive techniques (e.g. by systematic parameter variation) on workstations are becoming the means of developing suitable grids. The result is judged by the human eye; it is therefore essential that the 3D mesh is visualized and studied from all view angles. The grid has to be rotated and translated at will, and it should be possible to display and zoom in on hidden lines or subsets of the display, so that critical regions in the mesh can be detected and adjusted. It should be stressed that human intervention is unavoidable and at present no grid generation package is general enough to construct "good" meshes without some input from the expert. In Figure 11.1, the first and last phase of such a grid generation process is shown for a deformed square. Although the mesh on the left is regular and may seem acceptable, careful observation shows 'overspill' near the left hand corner. In addition, physical considerations (e.g. important viscous effects near solid boundaries) impose thin grid elements at the boundaries themselves, and this was achieved with the grid on the right. Given the geometrical simplicity and the two-dimensionality of the domain under consideration the iterative process required only a few steps for a satisfactory result to be obtained, but for complex geometries many iterations may be needed.

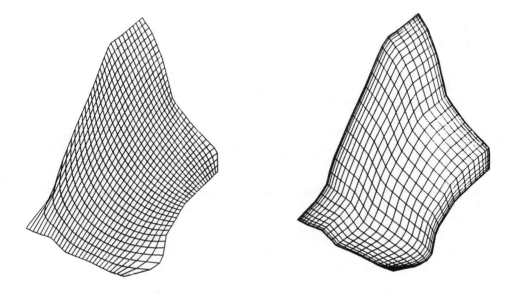

Figure 11.1. First and last phase of the mesh generation for a deformed square

At present graphics are not widely used in monitoring the solution process of fluid dynamics problems, but it is expected that within five to ten years this practice will become common.

For the analysis of the converged numerical solutions, the use of graphics is indispensable. However, human intervention is still essential. An unthoughtful representation of velocity vectors and pressure is unlikely to be effective in conveying all the information that is contained in the data of a complex 3D computation. One needs to have beforehand some idea of what quantities to display and in which part of the domain to look. When solving the Navier-Stokes equations, the vorticity plays an important role. It is important to know where it is generated and where it is transported. Contour lines of vorticity are a good indication of the spatial accuracy of the solution, and the presence of "wiggles" may suggest that mesh refinement is needed. Particle traces can be used to give an indication of the nature of the flow, and to pinpoint regions of flow reversal, vortices, etc. For some applications, the pressure field is used as the most sensitive indicator. In water turbines, for example, the low pressure regions are of interest because cavitation is likely to occur there (a phenomenon which causes erosion and eventually destruction of turbine blades), whereas for supersonic flows sharp gradients in the pressure field may indicate the presence of shock waves.

Visualization is, in this phase, first used as a quality control instrument, i.e., to provide the experienced CFD analyst with the information necessary to judge the quality of the solution. If the solution is found acceptable, visualization is used to obtain a better understanding of the physics, and/or to detect critical flow regions due to the geometry of the structure under study.

11.3. A graphical package for fluid flow visualization: CACTUS

To design a graphical package to serve the widest possible user base means finding a compromise between the following three conflicting requirements:

- provide the largest possible range of options,
- make it easy to use, and
- minimize the processing times.

Achieving the first two aims depends largely on the skills of the software writer, whereas the fulfillment of the third is constrained by the processing speed of the available hardware which, despite the recent technological advances, still does not satisfy present demands.

CACTUS is an example of a package which was written with the flow simulation user in mind. It can be utilized to produce 3D or 2D scatter, line, surface and vector plots of multiblock data with cubical topology. Its internal processing capabilities allow for the calculation and display of contour lines and iso-surfaces. An additional pre-processing package compatible with CACTUS can be used for the calculation of streamlines or stream-surfaces. This range of options has been found to cover a wide range of user needs.

The user is able to alter and store most of the display settings. Thus, he can customize his working environment. A sensible set of defaults is always available to those who do not want to explore the options available. It is possible to change the line thickness, the color scales, the vector length, etc. Very often several lines or surfaces are to be drawn, and

CACTUS allows the definition of 'types' of lines or surfaces (comparable to a *font* on a text processor), to be able to visualize different sets of data on the same display. It also permits the user to translate, rotate and to zoom in on the displayed results, to change the light source, etc.

With all these options, it is the user who finally selects how to display his results. Much attention has been paid to the structure of the command interpreter of the graphics software, in order to ensure that interactive use is fast and easy to learn. Devices such as the mouse, the button box and the knob box may also be used, see example in Figure 11.2 (color section). The rotation of a displayed image about one of its axis can be effected by rotating one of the knobs, which are supplied with the station hardware, and changing the light source on the image is done by pressing a button. CACTUS can also be used in batch, since it is able to read and execute shell scripts. A pre-stored set of images can be read and displayed on the screen in sequence.

11.3.1 Possible improvements in visualization

Two main sources of errors are encountered when visualizing data. The first is due to the limitations that arise when one tries to produce a 2D image that should be perceived as being in 3D space. Surface curvatures or the direction of vectors are, for example, difficult to recognize. Secondly, not all information to be displayed is continuous, which may pose problems. A typical example is the calculation of the iso-lines on the surface of a grid.

The present generation of graphical packages offers several aids, for example shading or depth-cuing (changing the intensities of fore and background colors), to improve the perception of the curvature of a line or surface. The same applies when displaying the direction of a vector. The graphical package CACTUS offers in this respect several possible representations of vectors, see also Figure 11.3 (color section). While the representation on the left is conventional, the one on the right provides insight into the direction of the vector. A cylinder of length proportional to the magnitude of the vector is used, connected to a conical tip of constant length. The ratio of the cylinder length to the cone length provides the information about the magnitude; the projection of the vector with respect to the reference coordinate system gives the direction. Finally, the two ends of the cylinder are displayed in different colors to enhance the perception of the direction of the vector.

Problems due to the discrete nature of the data can not be avoided, and one has to be always aware of their presence. For example they appear when trying to calculate the isolines for the square shown in Figure 11.4 in the displaying of the line $(L + H)/2$. The result depends on the algorithm used to calculate the isolines, and when using a graphics software package it is important to know which algorithm is being used. In CACTUS, for instance, the square is first divided in four triangles with the center of the square as common point, the value of which is set to the average value of the four corners. Finding the isolines in a square yields one unique solution, and for the example shown in Figure 11.4 the isolines produced by CACTUS form a cross.

Other sources of error found in visualization are indicated by Buning (1988).

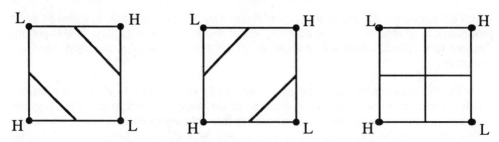

Figure 11.4. Possible isolines representation in an ambiguous situation

11.4. Selected applications

11.4.1 Inviscid hypersonic flow simulations

Stimulated by ambitious plans like the HERMES space shuttle, the German space plane Sänger and the American space plane NASP, hypersonic flows are now experiencing a rapid increase in interest. Hypersonic flows are high speed flows in which physical processes (chemical reactions, radiative heat transfer), usually neglected in low speed flows, are important. These physical processes are initiated by strong shock waves, which are generated in the flow field, and the correct modeling of these processes is vital in the design of vehicles flying at high speeds. As an example one may consider a hypersonic flow with a free stream Mach number of 25 over a blunt body. If the freestream temperature is 230 K, it can be found, using constant property perfect gas theory, that the stagnation temperature at the body would be 28,000 K. Correct modeling of the physical processes, however, yields stagnation temperatures between 6,000 and 9,000 K, depending on the freestream pressure. Experiments of hypersonic flows over the whole range of possible flight conditions are impossible due to the extremely high temperatures and pressures encountered which are difficult to reproduce experimentally. In the development program for the HERMES space shuttle numerical simulations are used to determine the aerodynamic coefficients for those flight conditions.

Below an altitude of 80 km hypersonic flows are described by the Navier-Stokes equations. In many applications viscous effects are confined to a small region close to the body, and thus the Euler equations may be used as an approximation throughout the flow domain These (partial differential) equations describe the conservation of mass, momentum and energy, and are closed by the algebraic equation of state. The chemical reactions may be modeled at different levels of complexity. If it is assumed that the flow is in chemical equilibrium, only the equation of state is modified, and an algebraic system of equations connecting the composition of air to the internal energy and density is added. If it is assumed that the flow is in chemical non-equilibrium, partial differential equations for the composition of air have to be added to the Euler or Navier-Stokes equations.

Flow visualization is used to determine critical regions in the design of hypersonic vehicles, for example the interaction of a shock wave with various vehicle geometries, the interaction between two shock waves, the effect of a flap deflection on the flow field, the presence of large vortices in the flow field, etc. Figure 11.5 (see color section) shows an example of a 3D Euler calculation carried out at IMHEF within the framework of the industrial Hermes project (Avions Marcel Dassault - Breguet Aviation, IMHEF-EPF

Lausanne, ABB Turbo Systems Ltd.-Baden and Eidgenossische Flugzeugwerke - Emmen). The Euler solver uses an unstructured mesh, and the results are visualized using the BASPL system (see Chapter by S. Merazzi and E. Bonomi in this volume). The figure shows the Mach number contours on the Hermes space shuttle and in three different planes normal to the body. Due to the large incidence, the shock wave at the upper body is weak and hardly visible. The interest of this calculation lies in the flow around the lee side of the shuttle; here the shock wave is located very close to the surface. Regions of mesh refinement are clearly visible, especially around the winglets. The unstructured grid used looks bad, but this is only due to the fact that plane cuts through tetrahedric cells have been taken.

11.4.2 Flow in hydraulic turbines

Hydraulic turbines are devices used in the conversion of the energy contained in a head of water into kinetic energy, which is subsequently transmitted to electric generators. A complete hydraulic assembly consists of a spiral case inlet, which channels the incoming water, a distributor, with usually two rows of blades that direct the flow to the runner (turbine) at a specific incidence angle, and a diffuser at the exit of the rotating runner through which the water flows to the lake or reservoir. Historically the design of all the components of an hydraulic assembly has been based on empirical correlations but in recent years the numerical simulation of flows in turbines has augmented the available knowledge. The use of potential flow calculation techniques provides simplicity and speed in the acquisition of results but more accurate methods based on rotational (and sometimes also viscous) equations are rapidly gaining ground. The goal is the design of the complete installation by numerical simulation. Essentially the process is one of optimization, using blade shape, thickness, etc. as parameters, with the aim to maximize the overall efficiency of the installation without having to build and test too many prototypes.

A common technique for the computation of the flow in a Francis runner is based on the numerical solution of the Euler equations. This approach has been shown to yield results which are in excellent agreement with experiments (see Sottas *et al.* (1990)). For some of the other device components viscous effects are important (in particular for the diffuser), and solution of the Navier-Stokes equations is required. Apart from quantitative comparisons of velocity and/or pressure distributions at specific locations, results of flow simulations in hydraulic installations usually focus on the display of isobaric surfaces on the pressure and suction side of blades, and global velocity vector fields. In Figures 11.5 and 11.6 (see color section) there are views of interblade volumes of a Francis runner with the calculation domain and its discrete mesh shown in the background of these figures. Computational economy is achieved by exploiting the symmetry of the runner and the use of periodic boundary conditions. These figures show the pressure distribution and the velocity field on the suction and pressure sides of the runner blades respectively. The following useful information can be deduced. On the corner between the blade trailing edge and the runner band on the suction side of the blade, a region of low pressure is found, which is a likely source of cavitation - a result that has now been confirmed by experiment. A weak vortical structure appears close to the blade leading edge on the pressure side; this unexpected result could not have been detected by experiments. Improvements in the present blade design can start with these considerations in mind. It is now becoming desirable to develop interactive techniques that allow for the modification of the blade shape (and consequently of the mesh) while the computation is in progress and intermediate results are visualized. This would greatly enhance the optimization process.

11.4.3 Simulation of turbulence

Turbulence is the most common state of fluid motion and examples of it can easily be found in everyday life, as in a fast flowing river or in a rising plume of smoke from a chimney. For some technological applications, lack of understanding of the nature and effects of turbulent flows limits the ability of engineers to meet design goals.

While the equations that govern turbulent motion have been known for over a hundred years, there is still no understanding of the dominant dynamical processes that underlie the complex and irregular behavior that is observed. Analytical work has not been able to overcome the mathematical difficulties encountered in dealing with the nonlinear physical laws of motion. Experiments have produced a wealth of data that have led to some insights, but in most cases they are not able to provide information with the sufficient detail and accuracy needed to test turbulence models. A substantial advance is the establishment of the existence of statistically significant "coherent" motions within the seemingly random background, thus opening the possibility of devising theories and models that are based on a limited set of basic turbulent structures. The information presently available about these structures is in many respects incomplete and important questions such as those on the events leading to their generation are still unanswered.

Owing to the substantial increase and wide availability of computer power, a new source of data on turbulence has now become available. The new approach is called direct or full numerical simulation and involves the computation of an approximate solution to the equations of fluid motion at a set of discrete points in space and time. No model is used. Its unique contribution to turbulence research lies in the ability of the method to provide almost complete information about the state of the simulated flow field at a given time and - if possible - the time history of it. It is thus possible to study the details of the flow structures in space and time, an option not available through experimental methods. However, the cost of such computations is very high. The three dimensional and unsteady nature of turbulent structures have to be accurately captured by the numerical simulation. This imposes very severe resolution requirements. At present the flows that are being studied are confined within geometrically simple boundaries and to low Reynolds numbers. Typically, the number of grid points that are employed is of the order of several million with computing times of several hundred CPU hours per simulation on the currently available supercomputers. The output of the computation is essentially the velocity vectors and pressures for every grid point and is usually stored on magnetic tape for later analysis. To reduce the external storage requirements, only a sample of the time history of the whole flow field is stored. Even after this reduction the databases that are generated are of the order of several hundred gigabytes.

11.4.3.1 Postprocessing

The greatest technical difficulties in the postprocessing phase of the work are associated with the management of the database and will not be dealt with here.

The analysis of the results is divided into several stages of increasing complexity. At the simplest level one calculates flow averages over the whole flow field and time. Quantities like the friction factor and mean turbulent kinetic energy fall into this category. Experimental data on these quantities is usually available. This makes possible a first check on the accuracy of the computed solution. At the next level one calculates the various moments of the flow variables or moments of derived quantities. These include mean flow velocities, their skewness and flatness, Reynolds stresses, vorticity and their higher moments. Because of the geometric simplicity of the flow domains of present simulations and the additional assumption of homogeneity along at least one cartesian direction, all flow moments depend at most on two independent variables. For their graphical display, line and contour plot graphics packages are quite adequate.

At present, the study of the structure of turbulence demands more from the fluid dynamisicist than from the visualization expert. An example of the amount of information generated by the simulations is given in Figure 8 which shows the instantaneous secondary velocity field on a plane normal to the streamwise direction from a simulation of turbulence through a straight square duct. This figure represents 1 /(2 x 768)th of the amount of independent information that is generated each time step of the simulation.

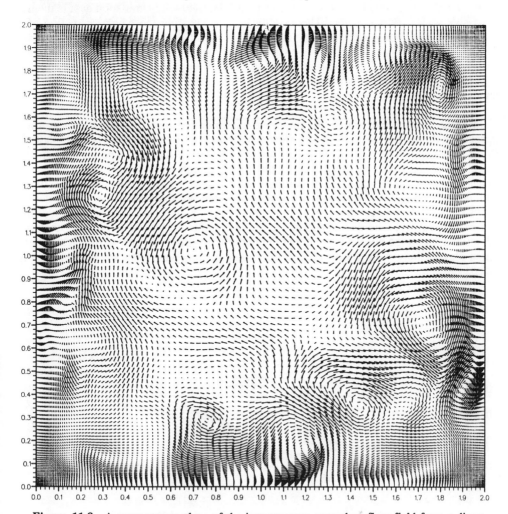

Figure 11.8. A cross stream plane of the instantaneous secondary flow field from a direct simulation of turbulence in square duct

Almost all of the ideas about the structures present in turbulent flows have originated either from correlation measurements, conditional sampling or from physical visualization techniques. However, these techniques give a rather limited picture of the 3D nature of the flow and their quantitative analysis is of limited accuracy. Similar visualization techniques can be employed in the simulated flow field by releasing "massless particles" into it. This technique has mostly confirmed the experimental observations but is not able to produce much more information. When using such markers to follow flow structures, it is found that the spatial definition of any discernable structure quickly degrades by the random transport of the outlining markers due to the background flow; a feature inherent to the physics of the flow. Additional problems arise from the fact that turbulent structures do not occur in identical copies but they are found to have a wide range of spatial scales, although they are topologically the same.

A related experimental technique which can also be used with simulation data is the use of passive scalar contaminants. This usually involves heating the whole or section of a boundary at low enough temperature for the buoyancy effects to be negligible. Contour plots of the temperature distribution at later times may indicate the extent and mode of mass and momentum transfer from the heated section. This, for example, may indicate that either large or small structures are the dominant feature and hence one may be able to deduce the length scales involved.

In order to track individual structures one has to devise a scheme for detecting their presence in the turbulent field. Here a great deal of uncertainty is involved because there is no a priori knowledge whether a given detection criterion has a unique relation to the "structure" being sought. Examples of such criteria are high vorticity regions for detecting high shear layers and low pressure regions for vortex detection. For such work the use of 2D contour plots or 3D isosurfaces of vorticity or pressure are quite sufficient. Figure 11.9 shows the instantaneous high cross-stream vorticity region on a plane normal to one of the four boundaries from the duct simulation. The use of the derived quantity rather than the velocity itself, brings out the presence of the underlying structures known as 'shear layers' which play an important role in generating turbulent energy but they are not completely understood.

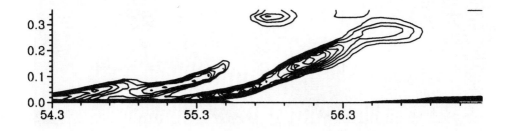

Figure 11.9. Instantaneous high cross-stream (into the page) vorticity region near one of walls of the simulated flow through a square duct. The mean flow is from left to right

The detection and visualization of turbulent structures at various realizations in simulated flows is only the starting point for understanding turbulence. In the second phase one has to determine the significance of their contribution in the mean characteristics of the flow. If a structure occurs very infrequently in the flow, it may have little impact on the flow statistics. Finally one has to determine their dynamical behavior. This is an area where animation of the flow structures may give an insight into the physics. Unfortunately present day computers are not sufficiently powerful for routine use of such an approach.

11.5 Closing remarks

At present, visualization is a common tool in the numerical simulation of flows and, as pointed out, it is of importance in different phases of such a simulation. Graphics are used to aid the CFD specialist in the understanding of complex flow phenomena. Up to now, it is always this specialist who has to interpret the results and translate them into something physically understandable.

For the near future, it is expected that the power (memory and speed) of workstations will increase further, making flow animations widely accessible. Moreover, interactive supercomputing will be used to optimize interactively the design of, among others, airplanes and hydraulic installations.

References

Buning, P.G. (1988). Sources of errors in the graphical analysis of CFD results. *Journal of Scientific Computing*, **3** (2), 149-164.

Buning, P.G. and Steger, J.L. (1985). Graphics and flow visualization in Computational Fluid Dynamics. In *AIAA 7th Computational Fluid Dynamics Conference*, Cincinnati, AIAA-85-1507-CP, pages 162-170.

Lasinski, T., Buning, P.G., Choi, D., Rogers, S., Bancroft, G. and Merritt, F. (1987). Flow visualization of CFD using graphics workstations. In *AIAA 8th Computational Fluid Dynamics Conference*, Honolulu, AIAA-87-1180-CP, pages 814-820.

Sottas, G., Drotz, A. and Ryhming, I.L., Eds. (1990). *3D Computations of Incompressible Internal Flows (a GAMM Workshop)*. To appear in Notes on Numerical Fluid Mechanics, Vieweg Verlag.

Thompson, J.F., Ed. (1982). *Numerical Grid Generation*, North-Holland.

Watson, V., Buning, P.G., Choi, D., Bancroft, G., Merritt, F. and Rogers, S. (1987). Use of computer graphics for visualization of flow fields. In *AIAA Aerospace Engineering Conference and Show*, Los Angeles.

12

Visualization and Manipulation of Medical Images

Osman Ratib
Hôpital Cantonal Universitaire de Genève

12.1. Introduction

During the last decade an increasing number of digital imaging techniques have been introduced in medical practice. Most radiologists and clinicians are now familiar with the usage and manipulation of digital images such as computed tomography scans (CT), magnetic resonance images (MRI) and nuclear medicine scans. Also with the development of high-capacity and high-performance computers, medical images that were traditionally recorded on films are now handled in digital form. Imaging modalities such as contrast angiography and ultrasound images are progressively shifting toward a fully digital form. Even conventional X-ray images can nowadays be recorded directly in digital form through photostimulable luminescence phosphore plates (Sonada, Takano et al. 1983). Another important factor for the development and wide spread of digital imaging techniques is the implementation of Picture Archiving and Communication Systems (PACS) in clinical environments (Huang, Mankovich et al. 1988). These systems provide a more efficient access to the medical images than conventional films.

The development of digital imaging techniques in medicine is accompanied by an increasing usage of image manipulation tools. These tools can be divided into two different categories: 1) general-purpose tools for image processing used to manipulate and modify the image presentation such as: intensity and contrast adjustment; image expansion reduction and rotation; filters for image smoothing and enhancement; and simple features extraction algorithms, and 2) more elaborate image analysis and measurement techniques for quantitative evaluation of the images. These tools are designed to assist the clinician in a more objective evaluation of the images and lead to a more refined diagnostic accuracy than visual interpretation alone.

12.2. Image Processing

The clinical use of digital image processing initially started with radionuclide studies in which digital images were recorded and stored in numerical form in computer memories. Therefore the first attempts to apply digital image processing techniques were directed toward the enhancement of radionuclide images addressing problems associated with their poor spatial resolution (Holman and Parker 1981). Later developments of CT and MRI scanners required further developments in mathematical algorithms for the reconstruction of tomographic images (sectional images) from three dimensional data sets.

Sophisticated tools are employed to generate the best possible result from the original data. Filtering and noise reduction techniques are widely used to enhance the signal to noise ratio (figure 12.1, see colour section). Conversely, edge enhancement techniques can provide sharper images with better perception of anatomical details. Filtering techniques can also be used to perform features extraction and image segmentation. A Laplacian filter for example can be used to generate "contours" around different components of an image based on their difference in intensity or brightness (figure 12.2).

Manipulation of image gray levels can also result in an improvement in image quality and a better perception of its content. Adjustable contrast and intensity levels can help differentiating between different anatomical structures. Non linear intensity transformations can improve the image contrast. Pseudocolor scales are also often used to enhance the visual perception of differences in intensity in an image. Nuclear medicine images are commonly visualized in pseudocolors to better differentiate areas with increased or decreased activity.

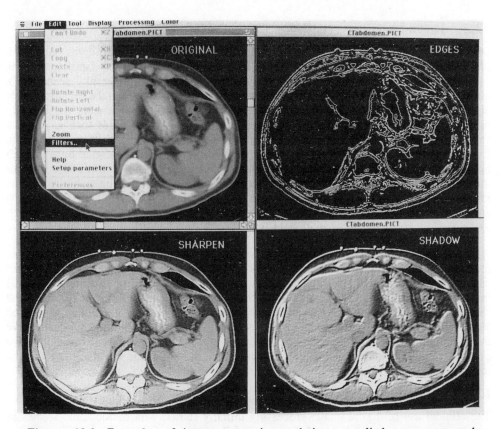

Figure 12.2. Examples of image processing techniques applied to a computed tomography image. Convolution filters were applied to the original image to sharpen the image (lower left), extract the edges (upper right) and enhance the texture of the organs by a shadow effect (lower right).

12.3. Quantitative Analysis techniques

With the advances in images acquisition and processing capabilities, many new analysis methods have been introduced, providing new approaches for quantitative assessment of medical images. The improvement in diagnostic efficiency due to these analysis techniques has been well demonstrated when they were compared to preexisting subjective and visual methods (Esser 1981; Ratib and Rutishauser 1988).

In cardiac diagnostic imaging for example a large number of numerical parameters which can be extracted from the images are clinically relevant. Almost all types of imaging modalities are clinically used in cardiology, and many of them measure the same parameters in different ways. Cine-angiograms, radionuclide ventriculography, ultrasound and MRI can all be used to assess ventricular wall motion for example. Coronary angiography, thallium scintigraphy and positron emission tomography (PET) are used to evaluate myocardial perfusion. Ultrasound, MRI, and fast CT are used to visualize cardiac malformations. Many similar quantitative measurement techniques can be applied to different imaging modalities. For example: 1) geometric estimates of ventricular size and performance using the Dodge elliptical model (Dodge, Sandler et al. 1960) can be applied to ultrasound, cineangiograms, radionuclide ventriculogram and MRI images, 2) blood flow and coronary reserve can be evaluated from contrast coronary angiography, conventional scintigraphy and PET using similar tracer kinetics models, 3) temporal behavior of regional ventricular wall motion can be analyzed from cineangiography, radionuclide ventriculography and MRI using factorial and Fourier analysis techniques.

12.3.1 Quantitative analysis of cardiac volumes

Computer analysis methods for the assessment of left ventricular (LV) function through the measurement of global parameters, such as volume and ejection fraction, have been widely accepted and are routinely used. Geometric and densitometric analysis techniques for the measurement of ventricular size and ejection fraction are applied to radionuclide as well as contrast ventriculograms, and echocardiograms (Collins and Skorton 1986).

Ventricular volume measurements play an important role in assessing cardiac function since ejection fraction and stroke volume can be calculated from the end-diastolic and end-systolic volumes. The most widely used method of measuring left ventricular volume is the area length method introduced by Dodge et al (Dodge, Sandler et al. 1960). This method models the left ventricle as a three dimensional ellipsoid. The three axes of the ellipsoid are estimated from the antero-posterior and lateral views of biplane left ventricular images. The same technique has been further extended to measurements made from a single projection. In the single plane method, it is assumed that the ellipsoid has rotational or circular symmetry about its long axis.

Alternatively, other techniques are based upon densitometry in which the densities or digital values of pixels within a region of interest are utilized. Densitometry utilizes the relative brightness values of picture elements (pixels) in regions of interest containing the left ventricular or the right ventricular area during systole and diastole to derive the ejection fraction. Only a single plane image is required to utilize this method. Furthermore, the technique is independent of left ventricular geometry. These densitometric measurement techniques can be applied to radionuclide angiograms as well as X-ray digital angiograms. In practice a number of technical problems make absolute densitometric measurements often difficult particularly with digital angiography where a number of correction factors must be introduced to compensate for non linear relationship between the image density and the corresponding blood volume.

12.3.2 Quantitative evaluation of vascular stenosis

Contrast angiography has been for many years the method of choice for the evaluation of vascular disease. This technique allows good visualization of vascular anatomy, but for better assessment of the functional significance of arterial stenosis and adequate management of patients there is a real need for more objective and quantitative evaluation of these lesions. With the recent development of digital angiography it has been possible to apply computer techniques for image processing and quantitative image analysis. Many methods have been proposed for the quantitative evaluation of the degree of stenosis using either densitometric approaches or geometric measurements (Rutishauser, Bussman et al. 1970; Smith, Sturm et al. 1973). Accurate measurement of the degree of stenosis remains, however, limited to lesions that are optimally imaged and, due to the low spatial resolution, is also limited to the large portions of the vessels.

Edge detection procedures to find the boundaries of the vessel can be implemented in a number of ways (Kirkeeide, Fung et al. 1982). Most techniques are based on one dimensional edge search procedures. The most common one is the gradient search method in which the first derivative of the density profile along each scan line is calculated (Wong, Kirkeeide et al. 1986). Whatever technique one uses for edge detection, calibration curves correlating the derived and the actual vessel edges must be developed with the help of phantom measurements. With the aid of calibration curves, edge detection accuracy down to 0.1 mm has been reported (Wong, Kirkeeide et al. 1986). Since the X-ray beam is divergent, the detected vessel edges must be corrected for both magnification and distortion effects to obtain absolute geometric measurements.

12.3.3 Analysis of cardiac scintigraphy

Single photon myocardial scintigraphy as well as positron emission tomography (PET) images of the heart provide images representing regional tissue tracer concentrations reflecting myocardial blood flow and metabolism. Quantitative analysis techniques for the measurement of relative tracer distribution as well as the measurement of absolute tracer concentrations and kinetics (from PET images) are required for a complete evaluation of these images. Conventional quantitative analysis of these images consisted of a manual selection of regions of interest to measure regional radiotracer activity. Manual analysis of these images is tedious and time consuming and a variety of automated analysis packages were developed for the quantitative analysis of cardiac tomograms (Ratib, Nienaber et al. 1988).

The particular shape of the tomographic images of the heart renders circumferential analysis techniques suitable for quantitative measurements of myocardial tracer tissue concentrations. Two different approaches are possible: a) the circumferential analysis based on the measurement of the myocardial activity in form of circumferential activity profiles, or b) a sectorial method based on the division of the myocardium into wedge shaped regions or sectors (Ratib, Nienaber et al. 1988). However a selection of a large number of sectors renders this approach similar to the circumferential analysis method (figure 3, see colour section).

12.4. Generation of parametric images

Quantitative evaluations of digital images often rely on synthetic images commonly referred to as functional or parametric images. In order to clarify the principles behind these methods some differences between the two types of images are briefly described here.

Functional imaging is the display of organ functions using images. The creation of functional images may be as simple as the imaging of a compound known to behave in a way which characterizes the function of an organ (i.e. when using radionuclide tracers in conventional nuclear medicine) or as complex as the result of a mathematical function derived from a set of images. The rationale for functional imaging (functional mapping or parametric imaging) is to allow the depiction of the status of a particular process within an organ. Complex mathematical procedures are used to localize and to highlight changes in a specific function of the organ that can't be visually detected directly from the original images. One can distinguish between direct functional imaging, in which images of the distribution of a particular tracer or dye are used as functional images, and indirect functional imaging, in which the functional images are formed by mathematical manipulations (Croft 1981).

Direct functional imaging: Conventional radionuclide images are the best example of direct functional imaging. Such images allow the localization of radiopharmaceuticals distribution corresponding to one or more physiologic processes in the explored organ. Similar approaches may be used in digital contrast angiography when the dye distribution is measured on the images and used as an index of the vascular flow. Videodensitometric measurement of cardiac volumes from contrast ventriculograms may also be considered as direct functional imaging since it does not require mathematical transformations of the original images into functional images.

Indirect functional imaging: Indirect functional imaging (more correctly called *parametric imaging*) is derived from mathematical manipulations of the original image data. Results of these calculations are reported as parametric images representing a topographic mapping of the calculated parameter. The purpose of the calculations is to detect and emphasize some property of the images which is usually not evident on visual inspection. The wide range of different kinds of parametric images may be separated into four basic types: 1) images formed from calculations on one image, 2) images formed from simple manoeuvres involving more than one image (addition, subtraction , division, etc...), 3) images formed from the application of mathematics to density vs. time curves for each point of the image, and 4) images formed from more complex mathematical functions.

12.4.1 Functional evaluation of dynamic cardiac images

Generating parametric images for the evaluation of regional cardiac function implies performing a data reduction and consequently isolating specific functional features from a cardiac cycle. Whether the latter has been obtained by a gated equilibrium blood pool image or by contrast cineangiography, is inconsequential. The data reduction generated by parametric imaging ultimately compresses all the information found in the cardiac cycle sequence - pertaining to a certain functional component - into one or two images. This makes for a much easier interpretation and singles out the respective functional component. Such images remove a great part of subjectivity in interpreting a cinematic display and enables instant comparisons at subsequent patient follow-up sessions. It is also important to remember that for parametric images, a color display is an absolute necessity and not a luxury. The most common parametric images obtained from gated radionuclide images are: the simple conventional function images such as stroke volume image, paradoxical image, and regional ejection fraction image as well as complex parametric images such as phase analysis images and time rate factor analysis.

Stroke Volume Image (SV): This image is simply obtained by subtracting pixel by pixel end-systolic from end-diastolic counts (ED-ES). The resulting image is a regional map of the net volume changes between diastole and systole in each pixel location (Pavel, Sychra et al. 1981). The display usually includes only positive values (negative values are set to 0), only those structures which contain more blood in end-diastole than in endsystole will be presented. Therefore, under normal circumstances only the ventricles will be seen.

Paradox Image: Originally applied in first pass and gated radionuclide studies this image is obtained by doing inverse pixel-by-pixel subtraction which means subtracting end-diastolic from end-systolic counts (ES-ED). The purpose of this image is to detect paradoxical ventricular wall motion. This is possible because dyskinetic regions will have increased local blood volume during systole.

Regional Ejection Fraction Image (REF): This image is obtained by dividing the SV image pixel-by-pixel by the background corrected end-diastolic frame. In this case, the use of background subtraction is mandatory. The regional ejection fraction image is proportional to local ejection fraction values. In normals, its pattern is represented by a number of crescent shaped areas of progressively lower value going from the free borders and apex to the septum and ventricular base. In regional wall motion abnormalities, the normal REF pattern is broken.

The clinical usefulness of global evaluation of ventricular function through the measurement of ventricular volume and ejection fraction is well demonstrated and widely accepted. However from a clinical point of view, a better evaluation of the ventricular function should allow the detection and the measurement of regional dysfunctions in ventricular wall motion (Daughter, Schwartzkopf et al. 1980). In order to discriminate between alterations in systolic and diastolic behavior, a quantitative analysis of regional ventricular wall motion should include a temporal analysis of the complete heart cycle. The three conventional parametric images described in the previous section have proven quite useful and have been utilized in several centers for several years. However, none of these images give information about synchrony of contraction. Of course, some indirect information exists in the sense that areas of abnormal wall motion have implicitly abnormal synchrony. Nevertheless, these images detect abnormal wall motion based on abnormal amplitude of contraction and not because of asynchrony. As a consequence, if the contraction amplitude is normal in a certain area, a co-existing asynchrony will be missed which occurs in most cases of electrophysiologic or ischemic disturbances. The two specific aspects mentioned above have been among the principle reasons for the success of the next category of images (Ratib and Righetti 1985).

Phase analysis of radionuclide angiograms: A technique based on pixel-by-pixel Fourier-phase analysis of radionuclide angiograms (Adam, Tartowska et al. 1979) was found to be very useful for the detection of regional asynchrony in wall motion. Phase and amplitude are obtained by what is commonly referred to as phase analysis, a more appropriate term could be "first harmonic analysis" while an inappropriate, yet often, used term is "Fourier analysis". The results of this analysis technique are displayed in color coded parametric images representing two dimensional maps of the sequence of contraction in the different regions of the heart. The amplitude image reflects the extent of contraction on a pixel-by-pixel basis. The phase image is nothing more than a topographical representation of relative values of timing of count rate changes. This in turn means that on a phase image, only the information pertaining to synchrony-asynchrony of motion is displayed. Being thus free of the amplitude characteristics, the phase image is in general no longer dependent on anatomy. Since 1979, phase imaging has been investigated for detection of ventricular wall motion abnormalities, evaluation of the cardiac electric activation sequence, and for many other applications (Ratib, Friedli et al. 1989).

Recently, phase analysis has been applied to variety of different imaging modalities and also to other organs than the heart. It has been used to study renal perfusion from radionuclide images, as well as to study lung perfusion patterns. Some investigators have also reported the applicability of this technique to dynamic CT and MRI images of the heart.

12.4.2 Dynamic analysis of blood flow

Using simple image manipulation techniques, the progression of contrast media through the coronary circulation can easily be visualized using parametric imaging techniques. Smith et al. (Smith, Frye et al. 1971) have described a method for the assessment of the arterial, myocardial and venous phases of contrast medium transit and the visualization of the spatial distribution and timing of contrast flow obtained from standard cine arteriograms. The results are displayed in a color coded parametric image where intensity and color modulation are used to depict the magnitude and timing of the appearance of contrast medium in each picture element respectively (figure 12.4, see colour section). Subsequent quantitative evaluation of changes in perfusion from baseline to hyperemic state provide useful information about the functional significance of coronary stenosis (Chappuis, Ratib et al. 1985).

12.4.3 Polar maps of tomoscintigraphic images

In order to summarize and facilitate visual interpretation of quantitative data derived from a series of contiguous cross-sectional images of the heart several parametric image display modes have been proposed (Esser 1981). The one that is most widely used for single photon tomographic images is referred to as a "Polar Map" or "Bull's Eye Display" (Caldwell, Williams et al. 1985). This technique displays data obtained from a series of cross-sectional planes in the form of concentric circles on one single parametric image (figure 12.5). Each slice is divided into an equal number of sectors, and the average or maximum counts per pixel are determined for each sector. These activity data are scaled to the sector with the highest activity in the entire left ventricular region of interest. The normalized activity for each slice is then reformatted and presented as a color coded map of concentric and progressively larger circles beginning with the most apical slice. By convention the central circle corresponds to the apex and the outermost circle to the base of the left ventricle. The angular orientation is maintained to match the heart orientation in the original images. In addition to a display of the relative activity distribution, the parametric images provide a convenient means for comparing two sets of images where differences in relative activity distributions over the entire heart can be easily evaluated by comparing the two corresponding polar maps. Further analysis of the difference between two polar maps can be obtained by taking either the difference or the ratio between the two and displaying the result in a third polar map (figure 12.6, see colour section). In this latter map it is possible to highlight regions that fall outside the difference or ratio limits established from studies in normals.

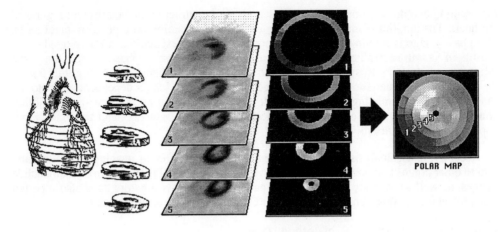

Figure 12.5. Generation of a polar map from a set of myocardial tomographic images. Regional tracer concentration measured in each image is displayed in a color-coded ring. Concentric rings are then displayed in a single parametric map.

12.5. Discussion and conclusions

The rapid evolution of workstations equipped with graphic user interface (GUI) will undoubtedly facilitate the user interaction and manipulation of complex analysis tools. A large number of workstation development projects reported in the literature tend to confirm this hypothesis (McNeill, Fajardo et al. 1987). Clinicians will in general benefit from the convenience and ease of use of a GUI. Most importantly, the graphic user interface has a significant impact on the way information is conveyed to the users. Proper utilization of a GUI allows presentation of data and information in a graphical or pictorial way that is more efficient than numerical or alphanumerical displays. Graphics and color coded maps can be used to present a very large number of data to the observer in a very rapid and efficient way. For example, graphic annotations and drawings overlaid on an image can provide a more efficient way of transmission of information to the user than written reports describing the features observed on the image.

Several of the specific cardiac analysis techniques described here were selected because they use visual and graphic displays to communicate a large amount of information. Several of them rely on color coded parametric images to display the results of complex image analysis. Functional and parametric images have an increasingly important role in the evaluation of radiological images and further developments in this area are expected due to the rapid improvement in hardware and software. Major efforts are currently being deployed in the design of new types of parametric images such as "Synoptic Images" and "Expert Images" (Pavel 1989).

Synoptic Images: There are routine clinical evaluations, such as gated exercise radionuclide angiographic studies, in which a large amount of quantitative data has to be interpreted. If there are on an average five levels of exercise (including rest and recovery) and if eight regional ejection fractions are considered for each level, it is easy to see that there is a need for data compression for synoptic (summarizing) interpretation of the whole study.

Such a synoptic image for gated exercise study has been proposed by Goris et al. (Goris and Briandet 1983).

Expert Images: In general the functional images describe the distribution of a single physical parameter related to function (ejection fraction, phase, amplitude, etc.) or a spatial distribution of a statistical measure. Such functional images have the following in common: A) they are based exclusively on input obtained from a particular patient; B) only a limited number of parameters (functional images) is used for the evaluation of each case; C) no direct statement about the diagnosis is made on these images.

As a consequence, the final diagnostic interpretation depends on the physicians experience and on his capacity to integrate the information provided by a constellation of functional images. There may also be more information in the dynamic sequence than is usually evaluated by the single parameter extracted in the parametric image. Recently, evidence has been forth coming that there exists an alternative, namely, the use of techniques available in the artificial intelligence domain, more specifically methods used in statistical pattern recognition (Sklansky and Wassel 1981). In essence, this means that the computer is first exposed to a "supervised learning" process by supplying a "training set" of cases and then asked to apply the new found knowledge to establish a certain diagnosis. The development of "knowledge based" analysis tools in conjunction with artificial intelligence algorithms will provide in the near future interactive computer programs that can assist clinicians and radiologists in the interpretation of diagnostic images. Programs capable of matching analysis findings with extensive databases of associations between radiological findings and differential diagnosis will allow the users the ability to rapidly formulate a set of likely differential diagnoses for a given constellation of radiological findings.

References

Adam, W. E., A. Tartowska, F. Bitter, M. Strauch and H. Geffers. (1979). Equilibrium gated radionuclide ventriculography. *Cardiovasc Intervent Radiol.* 2: 161-169.

Caldwell, J., D. Williams and J. Ritchie. (1985). Single Photon Emission Tomography: validation and application for myocardial perfusion imaging. In New Concepts in Cardiac Imaging. *Year Book Medical Publ.*: 115-136.

Chappuis, F., O. Ratib, B. Meier, A. Righetti and W. Rutishauser. (1985). Assessment of coronary flow reserve by computer analysis of digitized coronary angiograms at rest and after dipyridamol infusion. *J Am Coll Cardiol.* 2: 475.

Collins, M. C. and D. J. Skorton. (1986). *Cardiac Imaging and Image Processing.* McGraw-Hill.

Croft, B. (1981). Functional Imaging. *11th Symposium on the Sharing of Computer Programs and Technology in Nuclear Medicine.* 1-8.

Daughter, G. T., A. Schwartzkopf, C. Mead, E. Stinson, E. Alderman and N. Imgels. (1980). A Clinical Evaluation of Five Techniques for Left Ventricular Wall Motion Assessement. *IEEE conf. on Computers in Cardiology.* 249-252.

Dodge, H. T., H. Sandler, D. W. Ballew and J. D. Lord. (1960). The Use of Biplane Angiography for the Measurement of Left Ventricular Volume in Man. *Am Heart J.* (60): 762-773.

Esser, D. (1981). Functional Mapping of Organ systems and Other Computer Topics. *Proc. 11th Annual Symposium on the Sharing of Computer Programs and Technology in Nuclear Medicine.*

Goris, L. and A. Briandet. (1983). *A Clinical and Mathematical Introduction to Computer Processing of Scintigraphy Images*. New York, Raven Press.

Holman, B. L. and J. A. Parker. (1981). *Computer-Assisted Cardiac Nuclear Medicine*. Boston, Little Brown.

Huang, H., N. Mankovich, R. Taira, B. Stwart, B. Ho, K. Chan and Y. Ishimitsu. (1988). Picture Archiving and communication Systems (PACS) for Radiological Images: State of the Art. *CRC Critical Reviews in Diagnostic Imaging*. 28(4): 383-428.

Kirkeeide, L., P. Fung, W. Smalling and L. Gould. (1982). Automated Evaluation of Vessel Diameter from Arteriograms. (IEEE Computer Society): 215-218.

McNeill, K., L. Fajardo and B. Hunter. (1987). Design and Evaluation of Digital Image Workstations. (April): 31-33.

Pavel, D. G. (1989). Role of Functional Imaging in the Evaluation of Left Ventricular Function. 3, No. 2(April): 9-28.

Pavel, D. G., J. Sychra and E. Olea. (1981). Functional (Parametric) Imaging of Dynamic Cardiac Studies. *Effective Use of Computers in Nuclear Medicine*. NY, McGraw Hill:

Ratib, O., B. Friedli, A. Righetti and I. Oberhaensli. (1989). Radionuclide Evaluation of Right Ventricular Wall Motion After Surgery in Tetralogy of Fallot. *Pediatr Cardiol*. 10: 25-31.

Ratib, O., C. Nienaber, H. Shelbert, L. Bidaut, J. Krivokapich and M. Phelps. (1988). Automated Quantitative Analysis of Cardiac Positron Emision Tomography (Pet) Studies. *The Journal of Nuclear Medicine*. 29: 867-868.

Ratib, O. and A. Righetti. (1985). *Computer Analysis of Cardiac Wall Motion Asynchrony*. *Computer Generated Images*. New York, Springer-Verlag: 98-105.

Ratib, O. and W. Rutishauser. (1988). *Parametric Imaging in Cardiovascular Digital Angiography*. *Clinical Applications of Cardiac Digital Angiography*. NewYork, Raven Press: 239-251.

Rutishauser, W., W. Bussman, G. Noseda, W. Meier and J. Wellauer. (1970). Blood flow measurement through single coronary arteries by roentgendensitometry. Part I and Part II. *AJR*. 109: 12-34.

Sklansky, K. and N. Wassel. (1981). *Pattern Classifiers and Trainable Machines*. New York, Springer-Verlag.

Smith, H. C., R. L. Frye, D. E. Donald, G. D. Davis, J. R. Pluth, R. E. Sturm and E. H. Wood. (1971). Roentgen videodensitometric measurement of coronary blood flow. Determination from simultaneous indictor-dilution curves at selected sites in the coronary circulation and in coronary artery saphenous vein grafts. *Mayo Clin Proc*. 46: 800.

Smith, H. C., R. E. Sturm and E. H. Wood. (1973). Videodensitometric system for measurement of vessel blood flow, particularly in coronary arteries. *Am J Cardiol*. 32: 144-156.

Sonada, M., M. Takano, J. Miyahara and H. Kato. (1983). Computed radiography utilizing scanning laser stimulated luminescence. *Radiology*. 148: 833-838.

Wong, W., R. L. Kirkeeide and K. L. Gould. (1986). Computer Application in Angiography. *Cardiac Imaging and Image Processing*. McGraw-Hill: 206-239.

13

Visualization of Botanical Structures and Processes using Parametric L-systems

Przemyslaw Prusinkiewicz
and
Jim Hanan
University of Regina, Saskatchewan

Since their conception in 1968, Lindenmayer systems have played an increasingly important role as a tool for the description, analysis, developmental simulation and visualization of multicellular organisms. However, the range of applications was limited by the discrete nature of the formalism. This chapter describes an extension of L-systems which overcomes many of the earlier limitations. The key idea is to associate parameters with L-system symbols. The notion of a parametric L-system is defined formally, and its application to the modeling of biological phenomena is illustrated using a number of examples. The range of graphical applications of L-systems is extended to phyllotactic patterns and trees.
Key words: visualization of botanical processes, morphology and physiology of plants, realistic image synthesis, developmental model, L-system, turtle geometry.

13.1   Introduction

Visualization of biological structures and simulation of developmental processes are among the most exciting applications of computer graphics. The developmental processes can often be described using a small number of relatively simple rules. Graphical simulations make it possible to observe the effect of the repetitive application of these rules and present the results in an easy to understand form. The contrast between the simplicity of the descriptions and the complexity of the resulting structures cannot be fully appreciated without the assistance of a computer. In many cases, such as in the modeling of flowering plants, the final images are not only informative, but also aesthetically pleasing.

The rules that govern the development of plants can be captured in an elegant way using the formalism of Lindenmayer systems (Lindenmayer 1968). Although L-systems were originally designed to model development on the cellular level, their range of applications was soon extended to larger structures, such as inflorescences (Frijters and Lindenmayer 1974, 1976) and leaves (Lindenmayer 1977). However, only the topology of the modeled structures was defined formally. The geometry resulted from arbitrary decisions made by the draftsman.

The first set of rules for the automatic graphical interpretation of L-systems was given by Hogeweg and Hesper (1974), and refined by Smith (1984). These rules are *external* to the L-systems in the sense that no geometric information is included in the L-systems themselves. Instead, an L-system provides a topological description of the modeled structure, which is subsequently drawn using a set of interpretation rules applied uniformly to all structure components. External rules do not provide enough flexibility to visualize complex, differentiated structures, such as plants with leaves and flowers. To overcome this limitation, Prusinkiewicz (1986, 1987) proposed an alternative method for interpreting L-systems, related to earlier results of Szilard and Quinton (1979). A string generated by an L-system is considered as a sequence of commands for a LOGO-style turtle (Abelson and diSessa 1982, Papert 1980). The geometric information is *internal* to the L-system, that is, incorporated directly into the production rules. L-systems with turtle interpretation make it possible to synthesize realistic images of flowering plants (Hanan 1988, Prusinkiewicz *et al.* 1988, Prusinkiewicz and Hanan 1989). However, the modeling power of this technique is still limited. A major problem can be traced to the discrete nature of L-systems. In the practice of plant modeling, it commonly creates the following problems:

1. The unit angle by which the turtle can be rotated must be the smallest common denominator of all angle values in the structure. This means that many symbols may be required to represent larger angles. The same applies to the unit step size. Small twigs at the outer edges of a tree structure will determine the step size, although the main trunk could use a much larger unit.

2. The proliferation of geometric symbols means that the representation of a conceptual unit of a plant, such as an internode or branching point, may require an excessively large number of symbols rather than a single or few symbols.

3. Using a discrete model, it is difficult to capture continuous phenomena, such as the gradual elongation of stem sections and diffusion of substances in a growing structure. The obvious technique of discretizing the continuous values may require a large number of quantization levels, yielding L-systems with hundreds of symbols and productions.

Consequently, model specification becomes difficult, and the mathematical beauty of L-systems is lost.

In order to solve similar problems, Lindenmayer suggested that numerical parameters could be associated with L-system symbols (Lindenmayer 1974). He illustrated this idea by referring to the diffusion of nitrogen compounds in a filament, and to the continuous development of branching structures. Both problems were revisited in later papers (de Koster and Lindenmayer 1987, Janssen and Lindenmayer 1987).

This chapter introduces a formal definition of parametric L-systems, and uses them as a vehicle for surveying a wide range of graphical simulations of developmental processes.

13.2   Parametric L-systems

Parametric L-systems operate on *parametric words*, which are strings of *modules* consisting of *letters* with associated *parameters*. The letters belong to an *alphabet* V, and the parameters belong to the set of *real numbers* $\Re$. A module with letter $A \in V$ and parameters $a_1, a_2, ..., a_n \in \Re$ is denoted by $A(a_1, a_2, ..., a_n)$. Every module belongs to the set $M = V \times \Re^*$, where $\Re^*$ is the set of all finite sequences of parameters. The sets of all strings of modules and all nonempty strings of modules are denoted by $M^* = (V \times \Re^*)^*$ and $M^+ = (V \times \Re^*)^+$, respectively.

The real-valued *actual* parameters appearing in the words must be contrasted with the *formal* parameters used in the specification of L-system productions. Formal parameters may occur in logical and arithmetic expressions. If Σ is a set of formal parameters, then $C(\Sigma)$ denotes a logical expression with parameters from Σ, and $E(\Sigma)$ is an arithmetic expression with parameters from the same set. Both types of expression consist of formal parameters and numeric constants, combined using the arithmetic operators $+, -, *, /$, the exponentiation operator $\wedge$, the relational operators $<, >, =$, the logical operators $!, \&, |$ (not, and, or), and parentheses (). Standard rules for constructing syntactically correct expressions and for operator precedence are observed. Relational and logical expressions evaluate to zero for false and one for true. A logical statement specified as the empty string is assumed to have value one. The sets of all correctly constructed logical and arithmetic expressions with parameters from Σ are noted $\mathcal{C}(\Sigma)$ and $\mathcal{E}(\Sigma)$.

A *parametric OL-system* is defined as an ordered quadruplet $G = \langle V, \Sigma, \omega, P \rangle$, where:

- V is the *alphabet* of the system,

- Σ is the *set of formal parameters*,

- $\omega \in (V \times \Re^*)^+$ is a nonempty parametric word called the *axiom*, and

- $P \subset (V \times \Sigma^*) \times \mathcal{C}(\Sigma) \times (V \times \mathcal{E}(\Sigma))^*$ is a finite *set of productions*.

The symbols : and $\rightarrow$ are used to separate the three components of a production: the *predecessor*, the *condition* and the *successor*. For example,

$$A(t) : t > 5 \rightarrow B(t+1)CD(t \wedge 0.5, t-2) \qquad (13.1)$$

is a production with predecessor $A(t)$, condition $t > 5$ and successor

$$B(t+1)CD(t \wedge 0.5, t-2) \,.$$

A production *matches* a module in a parametric word if the following conditions are met:

- the letter in the module and the letter in the production predecessor are the same,

- the number of actual parameters in the module is equal to the number of formal parameters in the production predecessor, and

- the logical expression C evaluates to *true* if the actual parameter values are substituted for the formal parameters in the production.

A matching production can be *applied* to the module. The result is a string of modules specified by the production successor. The actual parameter values are substituted for the formal parameters according to their position during evaluation of arithmetic expressions. For example, production (13.1) above matches a module $A(9)$, since the letter A in the module is the same as in the production predecessor, there is one actual parameter in the module $A(9)$ and one formal parameter in the predecessor $A(t)$, and the logical expression $t > 5$ is true for $t = 9$. The result of the production application is a parametric word $B(10)CD(3,7)$.

If a module x produces a parametric word χ as the result of a production application in an L-system G, we write $x \mapsto \chi$. Given a parametric word $\mu = x_1 x_2...x_m$, we say that the word $\nu = \chi_1\chi_2...\chi_m$ is *directly derived* from (or *generated* by) μ and write $\mu \Longrightarrow \nu$ if and only if $x_i \mapsto \chi_i$ for all $i = 1, 2, ..., m$. A parametric word ν is generated by G in a *derivation of length* n if there exists a sequence of words $\mu_0, \mu_1, ..., \mu_n$ such that $\mu_0 = \omega$, $\mu_n = \nu$ and $\mu_0 \Longrightarrow \mu_1 \Longrightarrow ... \Longrightarrow \mu_n$.

An example of a parametric L-system is given below.

L-system 1: Sample parametric L-system

$$\omega : \quad B(2)A(4,4)$$

$p_1 :$	$A(a,b)$	$: b < 3$	$\rightarrow$	$A(a*2, a+b)$
$p_2 :$	$A(a,b)$	$: b >= 3$	$\rightarrow$	$B(a)A(a/b, 0)$
$p_3 :$	$B(a)$	$: a < 1$	$\rightarrow$	C
$p_4 :$	$B(a)$	$: a >= 1$	$\rightarrow$	$B(a-1)$

In this and all following L-systems it is assumed that a module is replaced by itself if no matching production is found in the set P. The words obtained in the first few derivation steps are shown in Figure 13.1.

Productions in OL-systems are *context-free*, i.e. applicable regardless of the context in which the predecessor appears. They are sufficient to model information transfer from the parent module to its descendants. However, a *context-sensitive* extension of OL-systems is necessary to model information exchange between neighboring modules. For example, in a *2L-system* the predecessor consists of three components, called the *left context*, the *strict predecessor* and the *right context*. By convention, they are separated using the symbols $<$ and $>$. In the parametric case, each component is a letter from the alphabet V with associated formal parameters from the set Σ. All formal parameters may appear in the condition and the production successor.

A sample context-sensitive production is given below:

$$A(a) < B(b) > C(c) : a + b + c > 10 \rightarrow E((a+b)/2)F((b+c)/2) \qquad (13.2)$$

Production (13.2) can be applied to the module $B(5)$ which appears in a string

$$\cdots A(4)B(5)C(6) \cdots \qquad (13.3)$$

since the sequence of letters A, B, C in the production and in parametric word (13.3) are the same, the numbers of formal parameters and actual parameters coincide, and the condition $4 + 5 + 6 > 10$ is true. As a result of the production application, the module $B(5)$ will be replaced by a pair of modules $E(4.5)F(5.5)$. Naturally, the modules $A(4)$ and $C(6)$ can be replaced by other productions in the same derivation step.

Productions in 2L-systems use context on both sides of the strict predecessor. 1L-systems can be considered as a special case of 2L-systems in which one-sided context is used. L-systems 9 and 10 in Section 13.4 contain examples of both production types.

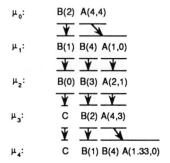

Figure 13.1: Strings generated by a parametric L-system.

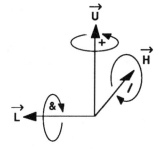

Figure 13.2: Controlling the turtle in three dimensions.

13.3   Interpretation

The interpretation of a parametric word proceeds as follows. After a string has been generated by an L-system, it is scanned from left to right and the consecutive symbols are interpreted as commands which maneuver a LOGO-like turtle in three dimensions. The turtle is represented by its *state* which consists of turtle *position* and *orientation* in the Cartesian coordinate system, as well as other attribute values, such as current *color* and *line width*. The orientation is defined by three vectors $\vec{H}, \vec{L}, \vec{U}$, indicating the turtle's *heading* and the directions to the *left* and *up* (Figure 13.2). These vectors have unit length, are perpendicular to each other, and satisfy the equation $\vec{H} \times \vec{L} = \vec{U}$. Rotations of the turtle are expressed by the equation:

$$\left[\begin{array}{ccc} \vec{H}' & \vec{L}' & \vec{U}' \end{array} \right] = \left[\begin{array}{ccc} \vec{H} & \vec{L} & \vec{U} \end{array} \right] \mathbf{R}$$

where $\mathbf{R}$ is a 3×3 rotation matrix. Changes in the turtle's state are caused by interpretation of specific symbols, each of which may be followed by parameters. If one or more parameters are present, the value of the first parameter affects the turtle's state. If the symbol is not followed by any parameter, default values specified outside the L-system are used as in non-parametric L-systems (Prusinkiewicz 1986, 1987). The following table lists the basic set of symbols interpreted by the turtle.

F(a)   Move forward a step of length $a > 0$. The position of the turtle changes to (x', y', z'), where $x' = x + a\vec{H}_x$, $y' = y + a\vec{H}_y$, and $z' = z + a\vec{H}_z$. A line segment is drawn between points (x, y, z) and (x', y', z').

f(a)   Move forward a step of length **a** without drawing a line.

+(a)   Rotate around $\vec{U}$ by an angle of **a** degrees. If **a** is positive, the turtle is turned to the left and if **a** is negative, the turn is to the right.

&(a)   Rotate around $\vec{L}$ by an angle of **a** degrees. If **a** is positive, the turtle is pitched down and if **a** is negative, the turtle is pitched up.

/(a)   Rotate around $\vec{H}$ by an angle of **a** degrees. If **a** is positive, the turtle is rolled to the right and if **a** is negative, it is rolled to the left.

[Initiate a branch by pushing the current state of the turtle onto a pushdown stack.

]   End a branch by popping a state from the stack and making it the current state of the turtle. No line is drawn, although in general the position of the turtle changes.

{   Start saving subsequent turtle positions marked by dots (.) as the vertices of a polygon to be filled.

.   Save the current position of the turtle as a vertex in the current polygon.

}   Fill the saved polygon using the current color index and line width.

~ X(a)   Draw the surface identified by the symbol **X**, scaled by **a**, at the turtle's current location and orientation. Such surfaces are usually defined using cubic patches (Hanan 1988, Prusinkiewicz 1987).

;(a)   Set the value of the current index into the color map to **a**.

#(a)   Set the value of the current line width to **a**.

It should be noted that symbols such as +, &, and / are used both as letters of the alphabet V and as operators in logical and arithmetic expressions. Their meaning depends on the context in which they are used.

Generation and interpretation of strings are not entirely independent. Their relationship is apparent when context-sensitive productions are used to generate branching structures (e.g. L-system 9 in Section 13.4.) In this case, modules representing adjacent tree segments may be separated by subwords which represent branches and are enclosed between square brackets. The context-matching procedure must skip over such subwords. For example, in the word

$$\text{ABCD[EF][G[HI[JKL]M]NOPQ]R}$$

D is the left context of G and N is the right context of G.

13.4   Visualization Examples

This section describes applications of parametric L-systems to the visualization of
processes and structures in botany. The specification of parametric L-systems adheres
to the notation introduced in Section 13.2. Furthermore, the following statements are
used:

- `#define` statements assign values to constants,

- `#include` statements list bi-cubic surfaces incorporated into the model,

- `#ignore` statements list modules which should not affect information transfer
  throughout the modeled structure, and therefore should be ignored while context
  matching.

Symbols `/*` and `*/` enclose comments.

The descriptions start from a botanical exposition of the problem, followed by a
presentation of the L-system and a discussion of the role of parameters in the model.

13.4.1   Spiral phyllotaxis — the cylindrical model

The term *phyllotaxis* literally means leaf arrangement, but it is used in a broader
sense to denote the regular arrangement of any lateral organs in plants (Erickson
1983). Spiral phyllotactic patterns evident in elongated organs can be described by
models which position components along a helix on the surface of a cylinder. These
patterns are generally characterized by the following formulae:

$$\phi = n * \alpha \qquad r = const \qquad H = n * h, \tag{13.4}$$

where:

- n is the ordering number of a component, counting from the bottom of the
  cylinder,

- ϕ, r and H are the cylindrical coordinates of component n,

- α is a constant *divergence* angle (Honda 1971) between the position vectors of
  two consecutive components, and

- h is the vertical displacement between two consecutive components, measured
  along the main axis of the cylinder.

An extensive mathematical theory developed by van Iterson (1907) and Erickson
(1983) describes the relationships between the values of r, α, and h corresponding
to various classes of phyllotactic patterns. Parametric L-systems make it possible to
generate these patterns without resorting to special-purpose programs, such as those
previously used by Fowler *et al.* (1989). For example, L-system 2 generates the spiral
pattern of disk-shaped components shown in Figure 13.3 (see color section).

L-system 2:   A cylindrical model of phyllotaxis

```
#define a 137.5    /* divergence angle */
#define h 35.3     /* vertical displacement */
#define r 500      /* component offset from the main axis */
#define s 50       /* scaling factor */
#include D         /* disk shape specification */
```

ω :   **A**

p_1 :   **A**   :   *   →   [&(90)f(r)~D(s)]f(h)/(a)A

The operation of this L-system simulates the natural process of *subapical develop-ment* characterized by sequential production of consecutive modules by the top part of the growing plant or organ. The apex **A** produces internodes **f(h)** along the main axis of the modeled structure. Associated with each internode is a disk ~ **D(s)** placed at a distance **r** from the axis. This offset is achieved by moving the disk away from the axis using the module **f(r)**, positioned at a right angle with respect to the axis by module **&(90)**. The spiral disk distribution is due to the module **/(a)** which rotates the apex around its own axis by the divergence angle **a** in each derivation step.

A realistic model of the sedge *Carex laevigata*, which displays spiral arrangement of spikelets in its inflorescences (spikes), is shown in Figure 13.4 (see color section). The L-system for the male spike (magnified on the left side of this Figure) is given below.

L-system 3:   Male spike of Carex laevigata

```
#define IRATE 1.025     /* internode growth rate */
#define SRATE 1.02      /* spikelet growth rate */
#include M              /* spikelet shape specification */
```

ω :   **A**

p_1 :   **A**   :   *   →   [&(5)f(1) ~ M(1)]F(.2)/(137.5)A

p_2 :   **M(s)**   : $s < 3$ →   M(s * SRATE)

p_3 :   **f(s)**   : $s < 3$ →   f(s * SRATE)

p_4 :   **&(s)**   : $s < 15$ →   &(s * SRATE)

p_5 :   **F(i)**   : $i < 1$ →   F(i * IRATE)

Production p_1 specifies the basic layout of the spikelets, similar to that given by L-system 2. The top portion of the spike has a conical shape due to the growth of the spikelets for some time after their creation by the apex. According to production p_2, a spikelet grows by factor **SRATE** in each derivation step, until it reaches the threshold size of 3. In an analogous way, productions p_3, p_4, and p_5 capture the distance increase between spikelets and the spike axis, the increase of the branching angle, and the elongation of internodes.

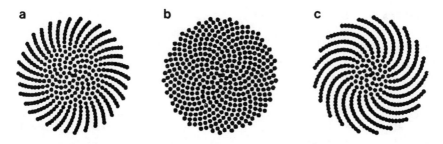

Figure 13.5: Generating phyllotactic patterns on a disk. These three patterns differ only by the value of the divergence angle a, equal to (a) 137.6°, (b) 137.5° (the correct value), and (c) 137.3°.

13.4.2 Spiral phyllotaxis — the planar model

The cylindrical model for phyllotaxis does not apply to flat structures, for example those found in many composite inflorescences. In order to describe them, Vogel (1979) proposed a planar phyllotaxis model expressed by the formulae:

$$\phi = n * \alpha \qquad\qquad r = c\sqrt{n},$$

where:

- n is the ordering number of a component, counting outward from the structure's center (the reverse order of floret age in a real plant),

- ϕ and r are the polar coordinates of component n,

- α is a constant divergence angle between the position vectors of two consecutive components (usually equal to 137.5°), and

- c is a scaling factor.

The above layout of components can be obtained in a straightforward way using the following L-system:

```
L-system 4:   A planar model of phyllotaxis

#define a 137.5    /* divergence angle */
#define s 50       /* scaling factor */
#include D         /* disk shape specification */

ω :   A(0)
p₁ :   A(n)  :  *   →   +(a)[f(n ∧ 0.5) ~ D(s)]A(n + 1)
```

Vogel's model is very sensitive to the value of the divergence angle a, as shown in Figure 13.5. An example of the application of Vogel's model to image synthesis is given in Figure 13.6 (see color section). The sunflower head was modeled by the following L-system:

```
L-system 5:   Sunflower head

#include S            /* seed shape */
#include R            /* ray floret shape */
#include M N O P      /* petal shapes */
```

ω : A(0)
p_1 : A(n) : $*$ $\rightarrow$ $+(137.5)[f(n \wedge 0.5)C(n)]A(n+1)$
p_2 : C(n) : n $<=$ 440 $\rightarrow$ $[;(1) \sim S]$
p_3 : C(n) : 440 $<$ n & n $<=$ 565 $\rightarrow$ $[;(2) \sim R]$
p_4 : C(n) : 565 $<$ n & n $<=$ 580 $\rightarrow$ $[;(3) \sim M]$
p_5 : C(n) : 580 $<$ n & n $<=$ 595 $\rightarrow$ $[;(4) \sim N]$
p_6 : C(n) : 595 $<$ n & n $<=$ 610 $\rightarrow$ $[;(5) \sim O]$
p_7 : C(n) : 610 $<$ n $\rightarrow$ $[;(6) \sim P]$

The layout of components is specified by production p_1, similar to that of L-system 4. Productions p_2 to p_7 determine colors and shapes of the components as a function of the derivation step number. The entire structure shown in Figure 13.6 was generated in 630 steps.

13.4.3   Trees

In spite of large interest in the modeling of trees for graphical purposes, L-systems have not been widely used for this purpose. One reason is that the tree structures, especially in moderate climates, strongly depend on environmental conditions, which are not directly captured by L-systems. In addition, even approximate models of tree architecture require precisely chosen branch growth rates and branching angles, which are difficult to express using nonparametric L-systems.

With the addition of parameters, L-systems become a convenient tool for specifying recursive algorithms for generating tree-like shapes, similar to those proposed by Oppenheimer (1986, 1989) and Bloomenthal (1985). An example is given below.

```
L-system 6:   A model for trees

#define DA 137.5    /* divergence angle */
#define BA 40       /* branching angle */
#define BR 1.732    /* branch width relative to mother branch */
#define LR 1.167    /* internode elongation rate */
#define WR 1.052    /* width growth rate */
```

ω : #(10)F(50)A(10)
p_1 : A(w) : $*$ $\rightarrow$ #(w)F(50)[&(BA)A(w/BR)]/(DA)[&(BA)A(w/BR)] /(DA)[&(BA)A(w/BR)]
p_2 : F(l) : $*$ $\rightarrow$ F(l $*$ LR)
p_3 : #(w) : $*$ $\rightarrow$ #(w $*$ WR)

Figure 13.7: Examples of trees.

The branching structure of the tree is defined by production p_1. In each derivation step, an apex $A(w)$ produces three new branches, each terminated by its own apex. The parameters DA and BA specify the branching angles as described by Honda (1971). The parameter w and constant BR relate the width of the mother branch to that of the daughter branches. Productions p_2 and p_3 capture the gradual elongation of branches and the increase of branch diameters over time.

Although parameter values yielding structures that suggest real trees can be obtained by trial-and-error, theoretical results may assist in proper parameter selection. According to Mandelbrot (1982, page 156), the ratio of branch diameters before and after bifurcation has been studied already by Leonardo da Vinci, who observed that "all the branches of a tree at every stage of its height when put together are equal in thickness to the trunk below them." In the case of the structure modeled by L-system 6, a mother branch with diameter d_1 gives rise to three daughter branches. Assuming that all of them have the same width d_2, we obtain $d_1^2 = 3d_2^2$, which yields the value of BR in L-system 6 equal to $\sqrt{3} \approx 1.732$. A general discussion of the relationships between the diameters of the mother and daughter branches, including alternative formulas to that proposed by da Vinci, is included in the book by MacDonald (1983, pages 131-135). Figure 13.7 presents three examples of trees generated using L-system 6. In addition to the constants in the L-system and the derivation length, the modified parameters include elasticity of branches and tropism direction (Prusinkiewicz *et al.* 1988, Prusinkiewicz and Hanan 1989). An extended version of this L-system, which produces leaves at the ends of branches, was used to generate the tree shown in Figure 13.8 (see color section).

13.4.4   Compound leaves

Herman, Lindenmayer and Rozenberg (1975) pointed out that "in the case of a compound leaf [...] some of the lobes (or leaflets), which are parts of a leaf at an advanced stage, have the same shape as the whole leaf at an earlier stage." The above observation is the cornerstone of the following L-system:

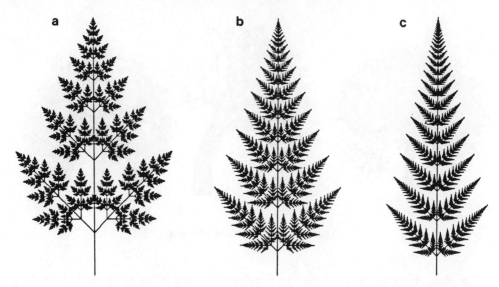

Figure 13.9: Examples of compound leaves.

L-system 7: A model of compound leaves

```
#define DELAY 1        /* apical delay */
#define RATE 1.5       /* internode elongation rate */
```

ω : A(0)
p_1 : A(d) : d > 0 $\rightarrow$ A(d − 1)
p_2 : A(d) : d = 0 $\rightarrow$ F(1)[+A(DELAY)][−A(DELAY)]F(1)A(0)
p_3 : F(1) : * $\rightarrow$ F(1 * RATE)

According to production p_2, in each derivation step the apex A(d) produces two segments F(1) and a pair of daughter branches. Production p_1 delays the development of the daughter branches by DELAY steps with respect to the mother branch. The branching pattern repeats recursively. Production p_3 gradually elongates the internodes, and in this way establishes proportions between leaf parts.

Sample leaves generated by the above L-system are shown in Figure 13.9. The corresponding parameter values are collected in the following table.

Figure	DELAY	RATE	Derivation length
a	1	1.50	16
b	4	1.23	30
c	7	1.17	45

The model is very sensitive to the growth rate values — a change of 0.01 visibly alters the spacing between leaf components.

13.4.5 Simple leaves

Although bracketed L-systems serve primarily as a method for generating branching
structures, they can be also applied to model surfaces such as leaf blades. One technique, described in (Prusinkiewicz *et al.* 1988) and (Prusinkiewicz and Hanan 1989),
creates leaf surfaces from polygons. Polygons are bounded by segments of the underlying branching structure and by extra edges which connect the terminal nodes.
The leaf shapes depend strongly on the relative growth rates and may change as the
segments elongate over time. An L-system modeling a class of simple leaves using this
approach is given below.

```
L-system 8:   A leaf model

#define LA 5            /* initial length - main segment */
#define RA 1            /* growth rate - main segment */
#define LB 1            /* initial length - lateral segment */
#define RB 1            /* growth rate - lateral segment */
#define PD 0            /* growth potential decrement */
```

ω : $\{.A(0)\#\}$
p_1 : $A(t)$:   * $\rightarrow$ $F(LA, RA)[-B(t).][A(t+1)][+B(t).]$
p_2 : $B(t)$: $t > 0$ $\rightarrow$ $F(LB, RB)B(t - PD)$
p_3 : $F(s, r)$:   * $\rightarrow$ $F(s * r, r)$

According to production p_1, in each derivation step the apex $A(t)$ extends the
main leaf axis by segment $F(LA, RA)$, and creates a pair of lateral apices $B(t)$. New
lateral segments are added by production p_2. The parameter t, assigned to the apices
$B(t)$ by production p_1, is decremented in each derivation step by the constant PD.
The production of new lateral segments stops when t reaches 0. Intuitively, the initial
value of the parameter t plays the role of "growth potential" of the branches. Segment
elongation is captured by production p_3.

It is convenient to divide a leaf modeled by L-system 8 into a basal area and an
apical area. In the basal area, the number of lateral segments is determined by the
initial value of growth potential t and the parameter PD. Since the growth potential
increases towards the leaf apex, the consecutive branches contain more and more
segments. In contrast, branches in the apical area exist for too short a time to reach
their full length. The actual number of segments there is proportional to the time
a branch had to grow, and gradually decreases towards the apex. As a result of
these opposite tendencies, a leaf reaches its maximum width near the middle of the
blade. For a given derivation length, the exact position of the branch with the largest
number of segments is determined by the value of the parameter PD. The following
table collects parameter values corresponding to the leaf models shown in Figure 13.10.

Figure	LA	RA	LB	RB	PD
a	5	1	1	1	0
b	5	1	1	1	1
c	5	1	.6	1.06	.25
d	5	1.2	10	1	.5
e	5	1.2	4	1.1	.25
f	5	1.1	1	1.2	1

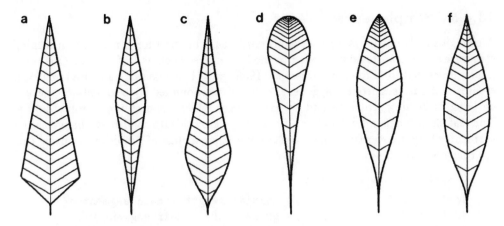

Figure 13.10: Examples of simple leaves.

If the parameter PD is equal to 0, all lateral branches have an unlimited growth potential, and the basal part of the leaf does not exist (a). If PD is equal to 1, the basal and apical parts contain equal numbers of lateral branches (b and f). Finer details of the leaf shape result from the growth rates. If the main axis segments and the lateral segments have the same growth rates (RA = RB), the edges of the apical part of the leaf are straight (a and b). If RA < RB, the segments along the main axis elongate at a slower rate than the lateral segments, resulting in the negative edge curvature of the apical leaf part (c). In the opposite case, with RA > RB, these edges have positive curvature (d, e and f). The curvature of the basal edges can be analyzed in a similar way.

13.4.6   Development of an inflorescence — *Mycelis muralis*

Graphical simulation of plant development employing various types of control mechanisms was discussed in Prusinkiewicz *et al.* (1988) and Prusinkiewicz and Hanan (1989), but some L-systems without parameters were too convoluted to be included there. We present here a developmental model of the inflorescence of *Mycelis muralis*, made presentable using a parametric L-system.

The development of *Mycelis* is difficult to model for two reasons. First, the plant exhibits a *basipetal* flowering sequence, which means that flowering starts at the top of the plant and proceeds downwards. Secondly, at some developmental stages the upper branches are more developed than the lower ones, resulting in an *acrotonic* plant structure. Both effects are in a sense counterintuitive, since it would seem that the older branches should start growing and producing flowers before the younger ones. To explain these effects, several models were proposed and formally analyzed by Janssen and Lindenmayer (1987). Below we restate their *Model II*.

L-system 9: Mycelis muralis

```
#include 0                    /* flower shape specification */
#ignore / + ~ 0
```

ω : I(20)FA(0)
p_1 : S < A(a) : * $\to$ T(0) ~ 0)
p_2 : A(a)   : a > 0 $\to$ A(a − 1)
p_3 : A(a)   : a = 0 $\to$ [+(30)G]F/(180)A(2)
p_4 : S < F   : * $\to$ FS
p_5 : F > T(a) : * $\to$ T(a + 1)FU(a − 1)
p_6 : U(a) < G : * $\to$ I(a)FA(2)
p_7 : I(a)   : a > 0 $\to$ I(a − 1)
p_8 : I(a)   : a = 0 $\to$ S
p_9 : S   : * $\to$ ϵ
p_{10} : T(a)   : * $\to$ ϵ

The axiom consists of three components. The modules F and A(0) represent an initial segment and the apex of the main axis. The module I(20) is the source of a signal representing a flower-inducing hormone or *florigen*. In nature, florigen is sent towards the apex by leaves located at the plant base.

The developmental process consists of two phases which take place along the main axis and are recursively repeated in branches of higher order. First, the main axis is formed in a process of subapical growth specified by production p_3. The apex produces consecutive segments F at the rate of one segment every three derivation steps (the delay is controlled by production p_2), and initiates branches G positioned at an angle of 30 degrees with respect to the main axis. (The symbol G is interpreted here in the same way as F.) At this stage, the branches do not develop further, simulating the effect of *apical dominance*, or the inhibition of branch development during the active production of new branches by the apex.

After a delay of 20 derivation steps, counted using production p_7, a flower-inducing signal S is sent *acropetally* (from the plant base towards the apex) by production p_8. Production p_4 transports S across the segments at the rate of one internode per step. Since new internodes are produced by the apex at a three times slower rate, the signal eventually reaches the apex. At this point, the second developmental phase starts. Production p_1 transforms the apex A(a) into a bud ~ 0. (Productions modeling subsequent phases of flower development are not shown.) Further branch production is stopped and a *basipetal* signal T(a) is sent towards the base in order to enable the development of lateral branches. The parameter a is incremented by production p_5 each time the signal T(a) traverses an internode. Subsequently, production p_6 introduces the value of the parameter a carried by the signal U(a) into the branches. The successor of production p_6 has the same format as the axiom, thus the module I(a) determines the delay between the initiation of branch development and the time signal S is sent to terminate further segment production. The delay a is the smallest for the top branches and increases towards the plant base. Consequently, the lower branches can grow longer than the higher ones. Thus, the parameter a represents growth potential of the branches, as previously discussed in the context of leaf models (L-system 8). Since the development of the upper branches starts relatively sooner, in some stages they will be comparatively more developed than the lower branches, and the flowering sequence will progress downwards, corresponding to observations of the real plant (Janssen and Lindenmayer 1987).

A developmental sequence of *Mycelis muralis* simulated using L-system 9 is represented in Figure 13.11 (see color section). Initially, the segments are shown as bright green. The passage of florigen S turns them pink, and the lifting of apical dominance changes their color to dark green. The model explains the flowering sequence and shape changes of *Mycelis*, and makes it possible to estimate the propagation rates of signals.

13.4.7   Development of a filament — *Anabaena catenula*

In de Koster and Lindenmayer (1987), the authors present a model which explains the pattern of cells observed in *Anabaena catenula* and other blue-green bacteria. The bacteria has the form of a non-branching *filament* and consists of two basic types of cells: *vegetative cells* and *heterocysts*. Usually, the vegetative cells divide and produce two daughter vegetative cells. In some cases the vegetative cells differentiate into heterocysts. The heterocyst distribution forms a well-defined pattern, characterized by a relatively constant number of vegetative cells separating consecutive heterocysts. The intriguing question is, how does the organism maintain constant spacing of heterocysts while growing.

The model explains the phenomenon using a well motivated hypothesis that the heterocyst distribution is regulated by nitrogen compounds produced by the heterocysts, transported from cell to cell across the filament, and decayed in the vegetative cells. If the compound concentration in a young vegetative cell falls below a specific level, this cell differentiates into a heterocyst.

A corresponding L-system is given below.

L-system 10:   A model of Anabaena catenula

```
#define CH 900              /* high concentration */
#define CT .4               /* concentration threshold */
#define ST 3.9              /* segment size threshold */
#include H                  /* heterocyst shape specification */
#ignore f ~ H
```

ω : $-(90)\text{F}(0, 0, \text{CH})\text{F}(4, 1, \text{CH})\text{F}(0, 0, \text{CH})$
p_1 : $\text{F}(s, t, c)$: $t = 1 \& s >= 6$ $\rightarrow$ $\text{F}(s/3 * 2, 2, c)\text{f}(1)\text{F}(s/3, 1, c)$
p_2 : $\text{F}(s, t, c)$: $t = 2 \& s >= 6$ $\rightarrow$ $\text{F}(s/3, 2, c)\text{f}(1)\text{F}(s/3 * 2, 1, c)$
p_3 : $\text{F}(h, i, k) < \text{F}(s, t, c) > \text{F}(o, p, r)$: $s > \text{ST}|c > \text{CT} \rightarrow$
                                     $\text{F}(s + .1, t, c + 0.25 * (k + r - 3 * c))$
p_4 : $\text{F}(h, i, k) < \text{F}(s, t, c) > \text{F}(o, p, r)$: $!(s > \text{ST}|c > \text{CT}) \rightarrow \text{F}(0, 0, \text{CH}) \sim \text{H}(1)$
p_5 : $\text{H}(s)$: $s < 3$ $\rightarrow$ $\text{H}(s * 1.1)$

Cells are represented by modules $\text{F}(s, t, c)$, where s stands for cell length, t is cell type (0 - heterocyst, 1 and 2 - vegetative types), and c represents the concentration of nitrogen compounds. Productions p_1 and p_2 describe divisions of the vegetative cells. They each create two daughter cells of unequal length. The difference between cells of type 1 and of type 2 lies in the ordering of the longer and shorter daughter cells. Production p_3 captures the processes of transportation and decay of the nitrogen compounds. Their concentration **c** is related to the concentration in the neighboring cells **k** and **r**, and changes according to the formula:

$$c' = 0.25(k + r - 3 * c) + c$$

Production p_4 describes differentiation of a vegetative cell into a heterocyst. The conditions specify that this process occurs when the cell length does not exceed the threshold value **st** = 3.9 (which means that the cell is young enough), and the concentration of the nitrogen compounds falls below the threshold value CT = .4. Production p_5 describes the subsequent growth of the heterocyst in size.

Snapshots of the developmental sequence of *Anabaena* are given in Figure 13.12 (see color section). The vegetative cells are shown as rectangles, colored accordingly to the concentration of the nitrogen compounds. The heterocysts are represented as red disks. The values of parameters CH, CT and ST were selected to provide correct distribution of the heterocysts, and correspond closely to the values reported in (de Koster and Lindenmayer 1987). The mathematical model makes it possible to estimate these parameters, although they are not directly observable.

13.5   Conclusions

This chapter formalizes the notion of parametric L-systems, and applies it to the modeling of structures and processes in plants. In L-system 2, the role of parameters is limited to the exact specification of constant angles and distances. L-systems 3 to 8 use parameters to simulate gradual changes of geometric features, such as internode lengths and branching angles. L-system 3 puts limits on these changes after final values have been reached. L-system 7 applies parameters to express developmental delay. In L-system 8, a variable parameter representing growth potential is used to limit the development of the lower parts of the leaf. A more complex example combining growth potential, delays and signals is represented by L-system 9. Finally, in L-system 10 parameters are used to represent continuous values of compound concentration.

In conclusion, a wide range of biological problems can be modeled, analyzed and visualized using L-systems. The introduction of parameters extends their modeling power.

Acknowledgements

Inspiring discussions with Professor Lindenmayer helped formalize the notion of parametric L-systems. Debbie Fowler and Dave Fracchia helped express the phyllotactic patterns and the model of *Anabaena* using parametric L-systems. Generous support from the Natural Sciences and Engineering Research Council of Canada, Apple Computer, Inc., and the Department of Computer Science, University of Regina, is gratefully acknowledged.

References

Abelson H. and diSessa A. A. (1982) *Turtle geometry.* M.I.T. Press, Cambridge.

Bloomenthal J. (1985) Modeling the mighty maple. *Computer Graphics*, 19(3):305–311.

de Koster C. G. and Lindenmayer A. (1987) Discrete and continuous models for heterocyst differentiation in growing filaments of blue-green bacteria. In *Acta Biotheoretica*, volume 36, pages 249–273. Kluwer Academic Publishers, The Netherlands.

Erickson R. O. (1983) The geometry of phyllotaxis. In J.E. Dale and M.L. Milthorphe, editors, *The growth and functioning of leaves*, pages 53–88. Cambridge University Press.

Fowler D. R., Hanan J., and Prusinkiewicz P. (1989) Modelling spiral phyllotaxis. *computers & graphics*, 13(3):291–296.

Frijters D. and Lindenmayer A. (1974) A model for the growth and flowering of Aster novae-angliae on the basis of table (1, 0) L-systems. In G. Rozenberg and A. Salomaa, editors, *L Systems*, pages 24–52. Springer-Verlag, Berlin. Lecture Notes in Computer Science 15.

Frijters D. and Lindenmayer A. (1976) Developmental descriptions of branching patterns with paracladial relationships. In A. Lindenmayer and G. Rozenberg, editors, *Automata, languages, development*, pages 57–73. North-Holland, Amsterdam.

Hanan J. S. (1988) PLANTWORKS: A software system for realistic plant modelling. Master's thesis, University of Regina.

Herman G. T., Lindenmayer A., and Rozenberg G. (1975) Description of developmental languages using recurrence systems. *Mathematical Systems Theory*, 8(4):316–341.

Hogeweg P. and Hesper B. (1974) A model study on biomorphological description. *Pattern Recognition*, 6:165–179.

Honda H. (1971) Description of the form of trees by the parameters of the tree-like body: Effects of the branching angle and the branch length on the shape of the tree-like body. *J. Theoretical Biology*, 31:331–338.

Janssen J. M. and Lindenmayer A. (1987) Models for the control of branch positions and flowering sequences of capitula in Mycelis muralis (L.) Dumont (Compositae). *New Phytologist*, 105:191–220.

Lindenmayer A. (1968) Mathematical models for cellular interaction in development, Parts I and II. *Journal of Theoretical Biology*, 18:280–315.

Lindenmayer A. (1974) Adding continuous components to L-systems. In G. Rozenberg and A. Salomaa, editors, *L Systems*, pages 53–68. Springer-Verlag, Berlin. Lecture Notes in Computer Science 15.

Lindenmayer A. (1977) Paracladial relationships in leaves. *Ber. Deutsch Bot. Ges. Bd.*, 90:287–301.

Macdonald N. (1983) *Trees and networks in biological models.* J. Wiley, New York.

Mandelbrot B. B. (1982) *The fractal geometry of nature.* W. H. Freeman, San Francisco.

Oppenheimer P. (1986) Real time design and animation of fractal plants and trees. *Computer Graphics*, 20(4):55–64.

Oppenheimer P. (1989) The artificial menagerie. In Christopher G. Langton, editor, *Artificial life: Proceedings of an interdisciplinary workshop on the synthesis and simulation of living systems*, volume VI of *Sante Fe Institute Studies in the Sciences of Complexity*, pages 251–274. Addison-Wesley Publishing Company, Redwood City, California.

Papert S. (1980) *Mindstorms: children, computers and powerful ideas.* Basic Books, New York.

Prusinkiewicz P. (1986) Graphical applications of L-systems. In *Proceedings of Graphics Interface '86 — Vision Interface '86*, pages 247–253.

Prusinkiewicz P. (1987) Applications of L-systems to computer imagery. In H. Ehrig, M. Nagl, A. Rosenfeld, and G. Rozenberg, editors, *Graph grammars and their application to computer science; Third International Workshop*, pages 534–548. Springer-Verlag, Berlin. Lecture Notes in Computer Science 291.

Prusinkiewicz P. and Hanan J. (1989) *Lindenmayer systems, fractals, and plants*, volume 79 of *Lecture Notes in Biomathematics*. Springer-Verlag, Berlin.

Prusinkiewicz P., Lindenmayer A., and Hanan J. (1988) Developmental models of herbaceous plants for computer imagery purposes. *Computer Graphics*, 22(4):141–150.

Smith A. R. (1984) Plants, fractals, and formal languages. *Computer Graphics*, 18(3):1–10.

Szilard A. L. and Quinton R. E. (1979) An interpretation for DOL systems by computer graphics. *The Science Terrapin*, 4:8–13.

van Iterson G. (1907) *Mathematische und mikroskopish-anatomische Studien uber Blattstellungen.* Gustav Fischer, Jena.

Vogel H. (1979) A better way to construct the sunflower head. *Mathematical Biosciences*, 44:179–189.

14

Molecular Computer Graphics and Visualization Techniques in Chemistry

Jacques Weber
and
Pierre-Yves Morgantini
University of Geneva, Switzerland

Abstract

Molecular graphics has become a major tool in computer-assisted chemistry, by allowing one to build, display and manipulate realistic models of the structure and properties of known and novel compounds as well. In this chapter, we will review the basic concepts of molecular graphics, ranging from the construction and representation of skeletal molecular models to the display of molecular surfaces color-coded according to a reactivity index. In addition, the techniques of visualization commonly used for the various chemical objects will be presented and discussed.

14.1 Introduction

Molecular graphics (MG), a recent technique of computer-assisted chemistry, can be defined as the application of computer graphics to investigate molecular structure, function and interaction. In view of spectacular progresses achieved in both computer graphics hardware and software, MG has known an important development in the last few years by providing tools for visualizing three-dimensional (3D) models of complex chemical compounds, such as pharmaceuticals and proteins, and their physico-chemical properties (Weber *et al*, 1989a). It is by no means an overstatement to assert that MG has become the indispensable complement of experimental chemical and biological tools, as exemplified by some applications where the contribution of MG is essential, such as drug design (Marshall, 1987), polymer and protein modeling (Karplus and McCammon, 1986) and materials science (Catlow, 1989).

A rapid access to the structural information concerning the compounds to be modelized is the very first step in any MG application. To this end, efficient procedures for structure and substructure searching within crystallographic data bases are commonly used (Mueller *et al*, 1989). When no experimental data can be found, theoretical model builders offering different degrees of accuracy are available in most molecular modeling packages (Frühbeis *et al*, 1987). This allows the user to rapidly generate computerized molecular models so as to interactively visualize and manipulate them at will. In order to also display molecular shapes and volumes, powerful algorithms have been suggested which generate molecular surfaces representative of the steric properties of the compounds under study (Connolly, 1983). Then, a large number of theoretical models are available to evaluate physico-chemical properties, such as electron densities, interaction potentials or super-delocalizabilites (Quarendon *et al*, 1984; Weber *et al*, 1988a), and to color the surfaces according to the property value used as a local reactivity index (Weber *et al*, 1989a). This leads to pseudo 4D-images conveying the maximum amount of information in view of the investigation of structure/activity relationships.

Reflecting the rather large range of chemical objects to be represented and manipulated, the techniques of visualization used in MG applications are numerous and diverse in complexity. First of all, the purely stereochemical aspects of molecular structures can be at best represented using skeletal (or stick) and ball-and-stick models. In order to emphasize the steric aspects of the structures, space-filling models are more adequate. When attempting to display intermolecular interactions or the accessibility to solvent molecules, surfaces or envelopes surrounding the structures can be generated using different algorithms. Finally, dynamic modeling is important to visualize molecular vibrations, rearrangements, reactions and fluctuations. The imagery of molecular graphics is therefore complex and a broad range of techniques of visualization have been developed on the graphics workstations currently used in MG applications.

In this chapter, we shall review the basic concepts of molecular graphics together with the techniques of visualization used in the following applications :

- representation, manipulation and comparison of molecular models;
- structural model building;
- molecular dynamics;

- construction and representation of molecular surfaces;
- calculation and display of molecular properties.

14.2 Applications

14.2.1 Representation and comparison of molecular models

Molecules are made of atoms linked together by chemical bonds. Beyond its chemical formula, an important piece of information concerning a given compound is therefore topological by nature, using a chromatic graph-like structure to describe the types of nodes (atoms) and the different connections (bonds). Figure 14.1 displays an example of such a chemical graph called developed formula, namely that of the cholesterol molecule (formula $C_{27}H_{46}O$), a steroid well known to be the cause of important health problems for a vast number of human beings !

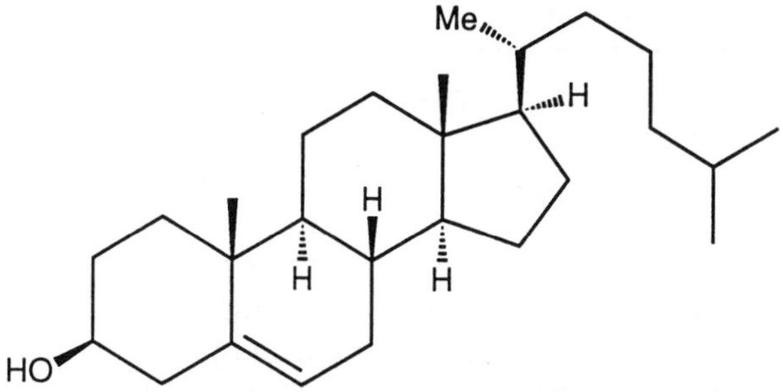

Figure 14.1. Developed formula of cholesterol, $C_{27}H_{46}O$

When coded in the coherent and compact form proposed by Dubois (1974), this topological pattern can be used for the management and processing of large computerized data files containing several millions of chemical compounds such as that of the Chemical Abstracts (Rhodes, 1985). However, such a coding is purely topological and, therefore, it does not contain the metric information which is indispensable for a 3D modelization of the chemical structures. It is then essential to have access to structural data files, obtained either from x-ray crystallography or from any other experimental technique (Allen *et al*, 1983), or to generate realistic 3D molecular architectures using ad hoc computer programs known as model builders (Burkert and Allinger, 1982). When this is achieved, it is possible to represent models of the structures on graphics displays using various techniques. The first one is the representation of the molecular skeleton as a stick model, with vectors depicting the various bonds (Fig. 14.2, see color section), the use of color-coded lines allowing one to differentiate the type of atoms.

The second technique leads to the so-called ball-and-stick model, with Phong or Gouraud shaded color spheres depicting atoms connected by solid cylinders (Fig. 14.3, see color section). Efficient algorithms allow one to solve rigorously the problem of intersection between spheres and cylinders so that realistic 3D images can be displayed and manipulated almost in real time for molecules containing up to 50-100 atoms on a workstation such as the Iris 4D-70GT from Silicon Graphics. Another common model is the space-filling or CPK one (Fig. 14.4, see color section), which displays a molecule as the union of large intersecting solid spheres with van der Waals radii, centered on the atoms (Max, 1984). Although in this model the details of both molecular topology and geometry are no longer visible, it is frequently used as it leads to a good perception of the overall shape and volume of the compound under study. In this respect, the comparison between Figs. 14.2-4 (see color section) is interesting as it reveals that different aspects of molecular structure are emphasized in the different models, which practically means that all of them should be available on graphics equipments used for MG purposes.

Another important point lies in the capability of the chemist to interact with the system: a beautiful image in the form of sophisticated 3D pictures would be of little help, if any, without the possibility to manipulate in real time or so the chemical models in order to visualize every structural detail, to superimpose molecular structures, etc. Here again, these operations have been considerably facilitated by the rate of hardware (and sometimes software) developments reached in the last few years, which allows one to consider them today as routine manipulations on most graphics systems. Figure 14.5 (see color section) illustrates a comparison of two steroids, the sexual hormones progesterone and testosterone, obtained by superimposing the structural skeletons to emphasize their topographical similarities, which is easily performed using the MOLCAD molecular modeling package (Brickmann, 1988). The superposition presented in Fig. 14.5 (see color section) shows that, in addition to the different substituents on the rightmost cycle, both compounds exhibit indeed slightly different conformations for the remaining portion of the steroids. This underlines the fact that small differences in the chemical formulae of two compounds may induce subtle conformational changes leading in turn to important modifications in biological activities. It is therefore essential in MG applications to have at hand software modules enabling one to generate realistic 3D architectures for compounds with given formulae and the basic principles of model builders will be briefly described.

14.2.2 Structural model builders

A very convenient procedure to optimize the 3D geometry of a compound is to minimize its molecular or steric energy calculated using highly parametrized force fields describing the major interactions between the

Figure 11.2. Example of the use of hardware devices (button and knob boxes) for the manipulation of displayed results.

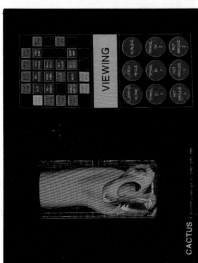

Figure 11.3. Two among the possible representations of velocity vectors that CACTUS allows.

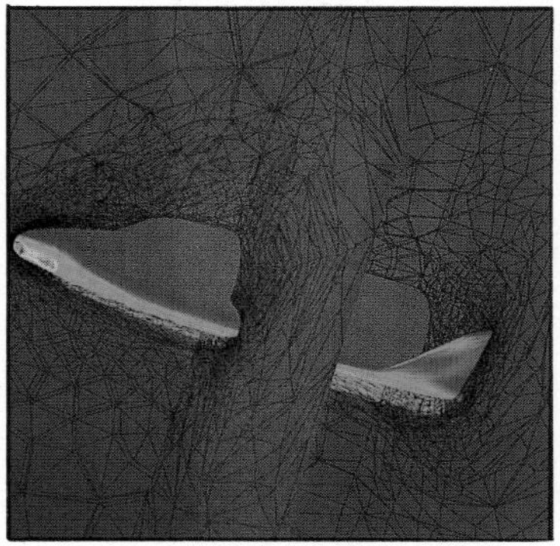

Figure 11.7. Isobars and velocity fields on pressure side of blades for Francis runner.

Figure 11.6. Isobars and velocity fields on suction side of blades for Francis runner.

Figure 11.5. Mach number contours for flow over the Hermes space shuttle.

B

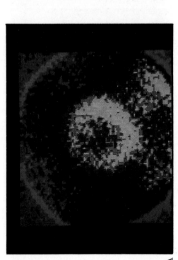

A

Figure 12.1. Illustration of image processing applied to radionuclide images of the heart. A) original image B) same image after interpolative background subtraction and smoothing.

Figure 12.3. Results of circumferential analysis of radionuclide images of the heart. Two images obtained at rest (lower left) and after exercise (lower right) are compared. The curves representing the regional tracer activity around the myocardium in these two images are shown on the upper left corner.

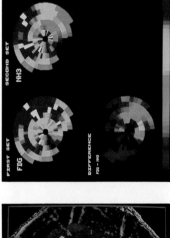

FIRST SET   SECOND SET
FDG   NH3

DIFFERENCE
NH - NO

Figure 12.6. Polar maps generated from two sets of tomoscintigraphic images of the heart. The first set of images was obtained after injection of F18-Deoxyglucose (FDG) used to visualize the regional metabolic rate of glucose in the myocardium. The polar map on the upper left corner is a summary of regional FDG distribution in 16 contiguous slices of the heart. A second set of images were obtained in the same patient after injection of N13-Ammonia (NH_3), a myocardial flow tracer, and is summarized in the polar map in the upper right corner. The difference between these two maps is shown in the lower left corner representing the areas with a mismatch between flow and glucose metabolism.

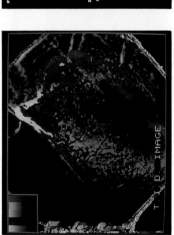

T I D I M A G E

Figure 12.4. Color coded parametric image generated from a sequence of coronary angiograms. The rate of progression of contrast media across the vascular tree is presented as a progression of colors. Areas with a reduced flow rate would appear in different colors.

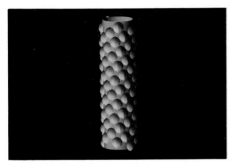

Figure 13.3. Model of phyllotaxis on the surface of a cylinder.

Figure 13.4. Model of the sedge *Carex laevigata*.

Figure 13.6. Model of a sunflower head.

Figure 13.8. A simple model of a tree with leaves.

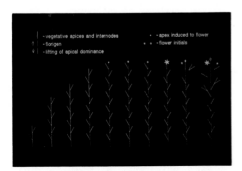

Figure 13.11. Developmental sequence of *Mycelis muralis*.

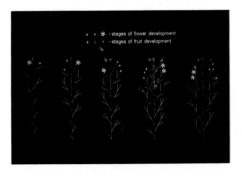

Figure 13.12. Developmental sequence of *Anabaena catenula*.

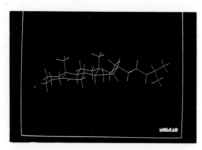

Figure 14.2. Stick model of cholesterol, represented on the IRIS 4D-70GTB using the MOLCAD package. Color coding: carbon: green; hydrogen: white; oxygen: red.

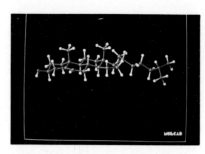

Figure 14.3. Ball-and-stick model of cholesterol.

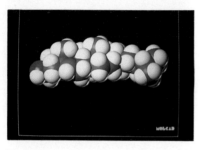

Figure 14.4. Space-filling model of cholesterol.

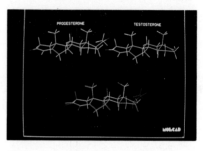

Figure 14.5. Comparison of the stick models built for the progesterone and testosterone steroids.

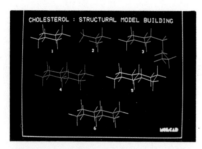

Figure 14.6. Several steps of building the model of cholesterol.

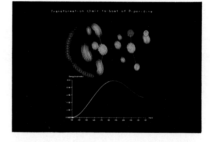

Figure 14.7. Interconversion path of the piperidine molecule from the chair to the boat form. Upper part: atomic trajectories represented in blurred motion on the Evans & Sutherland PS-390. Lower part: the associated energy curve. Color coding of the atoms: carbon: green; hydrogen: white; nitrogen; blue; oxygen: red.

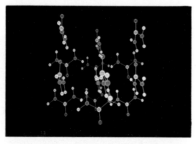

Figure 14.9. One normal mode of vibration of 8 unit cells of the urea crystal consisting of 16 molecules. The mode is an out-of-plane bending motion for one of the molecules in the unit cell and an in-plane rocking for the other.

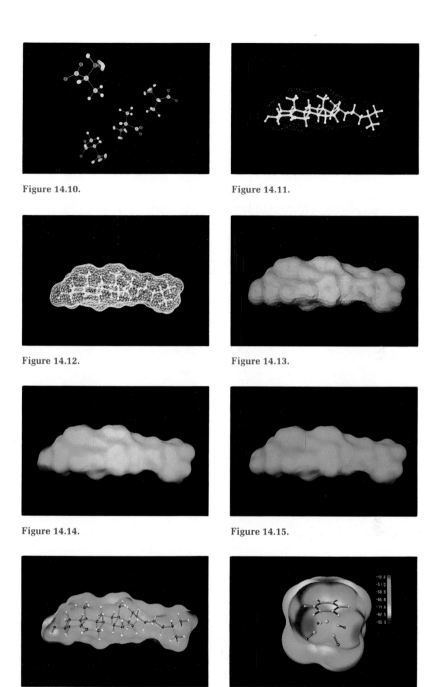

Figure 14.10.

Figure 14.11.

Figure 14.12.

Figure 14.13.

Figure 14.14.

Figure 14.15.

Figure 14.16.

Figure 14.17

Figure 14.10. A blurred picture of a normal mode of vibration of one unit cell (4 molecules) of the L-alanine crystal. The motion of this mode consists primarily of ammonium group H-N-H bending within the molecules. **Figure 14.11.** Dot molecular surface of cholesterol. **Figure 14.12.** Mesh molecular surface of cholesterol obtained from triangulation of the dot envelope. **Figure 14.13.** Faceted model of the molecular surface of cholesterol obtained from a solid rendering of triangulated surface. **Figure 14.14.** Smooth solid model of the molecular surface of cholesterol. **Figure 14.15.** Solid model of the molecular surface of cholesterol, with side clipping and partial representation of the structural model. **Figure 14.16.** Solid model of the electrostatic potential (MEP) of the cholesterol represented on the molecular surface. The most reactive site for the electrophilic attack corresponds to the red zone. **Figure 14.17** Solid model of the molecular surface of the $(C_6H_6)V(CO)_4{}^+$ complex colored according to the E_{int} property. The most reactive site for nucleophilic attack corresponds to the red zone. The color scale on the right indicates the E_{int} values mapping, in kcal/mol.

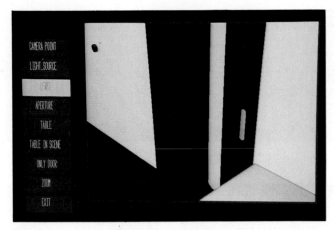

Figure 15.10. Though only 256 colors are available and no transparency and shadows are possible, quite realistic images of the scene can be obtained.

Figure 15.11. The door was successfully and easily found in the scene and correctly marked.

Figure 15.12. Samsung produced with Explore, TDI.

Figure 16.5.   Region matching using tree pruning.

Figure 16.3.   A stereo pair of an office and the corresponding segmentation results.

Figure 16.6.   Detection of "single" regions.

Figure 16.4.   Top-down segmentation results: top left image: left image of figure 3 in pseudo-colors; other images: various segmentation levels.

Figure 16.7.   Asymmetrical matching results (regions of figure 16.7 not existing in figure 16.6).

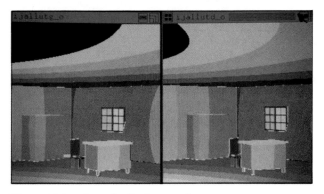

Figure 16.9.   A synthetic stereo pair.

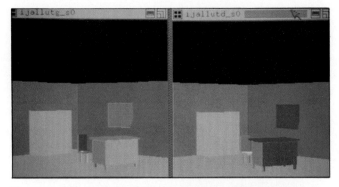

Figure 16.10. Results of the segmentation of figure 16.9.

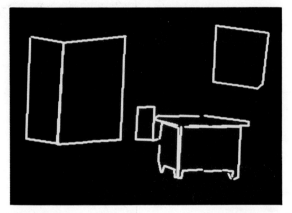

Figure 16.11. 3D reconstruction from the stereo pair of figure 16.10.

Figure 16.12. A ray traced version of an office (from Arnaldi et al. 1987).

nuclei and the effective potential generated by the electrons according to the laws of classical mechanics (Wilson, 1986). The empirical energy function derived from these force fields and which is generally used in such so-called molecular mechanics techniques has the following form :

$$E = E_b + E_\theta + E_\varphi + E_\omega + E_{vdw} + E_e + E_{hb} \tag{1}$$

where E_b and E_θ are bond stretching and bond angle bending energies, respectively, often assumed to be given by Hooke's law; E_φ is associated with the dihedral (torsion) energy, E_ω with the out-of-plane bending, E_{vdw} with non-bonded or van der Waals interactions, E_e with electrostatic interactions and E_{hb} with hydrogen bonding. Starting from a set of atomic coordinates generated from an interactive building of the structure using templates of fragments, the energy E is calculated in a first step. Then, the molecular geometry is systematically altered in order to find the lowest or global minimum by using robust gradient techniques (Mueller *et al*, 1989). However, as the number of degrees of freedom of the potential energy hypersurface is 3N-6, where N is the number of atoms, the problem of finding the global lowest minimum among numerous local ones is real. This explains why molecular dynamics simulations, where the atoms are allowed to move according to Newton's law in the molecular force field, are preferred for large molecules such as proteins (McCammon and Karplus, 1979). Combined with static molecular mechanics calculations, these simulations generally lead indeed to a realistic evaluation of the various energy minima on the potential energy surface, including the lowest one, provided they can be performed on a time scale large enough. This means that such calculations are not yet performed on a routine basis for proteins, as they are very expensive in terms of supercomputer time !

As a summary, model builders based on molecular mechanics are rapid and efficient due to the very simple and highly parametrized physical model used to calculate the energy components of a chemical system comprising up to several thousands of atoms. They are mainly used in organic and bio-organic chemistry, where several elaborate parametrizations have been suggested in the last few years (Burkert and Allinger, 1982; Rasmussen, 1985). As an example, Fig. 14.6 (see color section) displays several steps of the building of the structure of cholesterol using the molecular mechanics modules of the MOLCAD package. It is seen that the use of templates as model substructures is very useful as it allows the chemist to rapidly build realistic 3D molecular architectures for large systems, the molecular mechanics modules taking care of finding the optimal geometry for the connecting linkage between the various groups of atoms.

14.2.3 Molecular dynamics

Another recent and important application of molecular graphics consists in animating computerized models representative of dynamic chemical processes so as to simulate in vitro or in vivo situations. Indeed, as the collection of atoms constituting one or several molecular structures never remains rigid in normal conditions, real-time dynamic modeling is of prime utility. In this context, MG is an ideal tool through which to visualize the changes of a system as a function of time and a spectacular application may be found today in the dynamic modeling of the atomic trajectories along molecular rearrangements or chemical reaction paths (Weber *et al*, 1983 and 1987; Jefford *et al*, 1984). To this end, structural data bases made of several conformations of the molecular systems along the reaction path must first be calculated using theoretical, i.e. molecular mechanical or quantum chemical, models and the difficulties of such computations are real. However, such dynamic modelizations have considerable potential for understanding reaction mechanisms and for teaching complex rearrangements which are difficult to perceive using the standard techniques of conformational analysis. An example of such applications is found in Fig. 14.7 (see color section), which displays the trajectories of atoms of piperidine, a cyclic amine with $C_5H_{11}N$ formula, along its interconversion path from the chair conformation to a less stable intermediate known as the boat form (Fig. 14.8) (Weber *et al*, 1990).

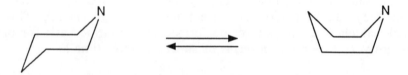

Figure 14.8. The chair (left) and boat (right) stable forms of piperidine

The procedure used to generate the animated sequences of this application is the following: the pre-calculated objects, which represent one single step of animation, are downloaded from the VAX-11/780 host computer into the local memory of the Evans & Sutherland PS-390 and stored in the display list by using the appropriate graphics support routines. To each of these steps is assigned one non-negative integer, and only the step whose number corresponds to the internal "level of detail" is displayed on the screen. By increasing (or decreasing) this level, one can switch from one step to another and therefore run through the whole sequence.

In another molecular dynamics application we have recently developed, an analysis of the normal modes of vibration of molecular crystals, known as harmonic dynamics, is performed using an empirical potential function model (Weber *et al*, 1990). The crystals chosen, namely urea and L-alanine

(Figs. 14.9 and 14.10, see color section) are important biomolecules of a rather small size and so can be used to test parameters for a protein or nucleic acid force field. Calculation and visualization of these vibrational modes is important as the atomic motions they involve help determine how a crystal will respond to external stresses or what reactions will occur within it. Such studies are very useful as tools for engineering crystals to exhibit specific properties or reactions (Wright, 1987).

14.2.4 Construction and representation of molecular surfaces

The concept of molecular surface has become increasingly important, as it contains an essential piece of information, namely the 3D shapes of the compounds, towards interpretation and prediction of molecular recognition and, ultimately, of chemical reactivity. Indeed, according to the well-known key and lock steric fit concept suggested as early as 1894 by Fischer (1894), the shapes of interacting partners should be complementary for a possible association and this condition is at best investigated and visualized by displaying their molecular surfaces.

Practically, the most common representation of the molecular surface on a graphic display is the dot model suggested by Connolly (1983), defined as the portion of the space-filling molecular model which is in contact with a probe sphere rolling on it and simulating a solvent molecule. This algorithm can also be used to calculate the solvent-accessible surface, which is traced out by the center of the probe sphere as it rolls over the molecule. The advantages of the display of dot surfaces on a real-time vector system are : (i) the transparency of the envelope, which allows one to simultaneously visualize the structural model, and (ii) the small number of points generated (typically 50-100 per atom) which enables the user to manipulate and clip the model in real time. Most of the molecular modeling packages available today in applications of computer-assisted chemistry allow one to generate with a very short response time the dot molecular surface of the compounds under study, as seen in Fig. 14.11 (see color section) which displays this envelope obtained for cholesterol. In macromolecular chemistry, dot surface models are very useful for localizing and visualizing in detail the binding sites of proteins, i.e. the active niches which are most likely to be attacked and blocked by an incoming reactant such as a drug molecule (Feldmann *et al*, 1978).

Another interesting representation of molecular surfaces is the 3D solid model, which is built from a mesh or chicken-wire envelope obtained from a triangulation of the dot surface. To this end, we have developed a triangulation algorithm similar to that suggested independently by Zauhar and Morgan (1988) and which can be briefly described as follows (Weber *et al*, 1988b). Using an input file made of the coordinates of the surface dots together with surface normal vectors, the algorithm is initialized by the connection of the first node to its nearest neighbor to form the first free edge (an edge is said to be free when the third point to form the triangle has not yet been selected). As every edge belongs to two triangles, the first edge is also the second free edge to be used for the construction of the opposite triangle. The algorithm then proceeds in the following way:

1. The next free edge is extracted from the list.
2. For that edge, the nearest node to its midpoint among those fulfilling the requested criteria (see below) is chosen to generate the triangle.
3. The previous operation leads to two new edges, each of them being either free, then added to the list, or already present in the list, where it will be labelled "complete".
4. The free edge of step 1 is labelled "complete".

Step 1 to 4 are repeated until no free edge is found.

The following criteria are applied for selecting a node so as to generate a triangle.

1. The new vertex must close the triangle in the same operational sense as all the previous triangles : in our case the clockwise direction as seen from outside has been chosen. This condition is tested via the sign of the triple product of edge, connection midpoint to vertex, normal vector.
2. The distance to the midpoint of the edge must lie within a given arbitrary range.
3. In order to generate a smooth surface, the angle between the connection midpoint to new vertex and the surface normal should lie within a given range around 90°.
4. None of the two new edges must intersect any other edge, which is tested (i) by checking if they are already present with the same operational sense in the list and (ii) by determining the intersections of all existing edges with the two planes formed by the new edges and surface normals.

This is a high-performance algorithm which leads to the results presented in Fig. 14.12 (see color section); it is frequently used in our laboratory as an intermediate step to represent molecular surfaces as 3D solid models, which are important in numerous applications. Starting from the polygons generated by the triangulation, it is indeed possible to represent a faceted model (Fig. 14.13, see color section) and a smooth 3D model obtained from it using rendering algorithms such as Phong or Gouraud shading, perspective setting and hidden surfaces treatment (Fig. 14.14, see color section). Clipping facilities are also available to simultaneously visualize the skeleton or ball-and-stick structural model (Fig. 14.15, see color section), which is indispensable to correlate the gross features of the surface with the structural characteristics.

The various models of molecular surfaces presented here are indispensable for an in-depth investigation of the 3D structures and their associated molecular shapes. However, in order to study fundamental problems in biological and chemical sciences such as molecular recognition (Lehn, 1988), it is essential to evaluate the most favorable site for the docking of reactants onto substrates in terms of interaction energies. To this end, molecular surfaces may be color coded to visualize various local physical properties such as electron densities, electrostatic potentials or intermolecular interaction energies to be used as reactivity indices, and this will be discussed in the next section.

14.2.5 Calculation and display of molecular properties

In addition to the display of the 3D shapes of chemical compounds, molecular surfaces provide an ideal support on which to map, using a standard color code, physical properties such as electrostatic potentials or interaction energies. Being pseudo 4D images, the models thus generated convey a very important additional information directly connected to the electronic structure of the compounds and which can be used as a local index of their reactivity. It is therefore not surprising that most molecular modeling packages offer the possibility to calculate and display such properties.

Historically, the molecular electrostatic potential (MEP) (Scrocco and Tomasi, 1978), defined as the electrostatic (Coulombic) interaction energy between a molecule and an external unit positive charge, has been the first property to be used as a reactivity index on molecular surfaces (Weiner *et al*, 1982). In this case, the Connolly surface was used and each dot colored according to the MEP value calculated at this point. On today's powerful workstations, however, 3D solid models are more convenient and they can be easily manipulated, which was only possible for dot surfaces a few years ago. The most reactive sites of a given compound are then immediately deduced from an inspection of the colored surface, as MEP minima (i.e. the most negative regions) correspond to the most favorable sites for an attack by a positively charged, or an electron acceptor, reactant whereas MEP maxima on the molecular surface are characteristic of regions easily attacked by a negative, or electron donor, reactant. In the applications we present here (see Fig. 14.16, color section) the color-coding range from red to yellow to blue extends smoothly over the numerical range of the MEP from the most negative to zero to the most positive values, which means that the red zones correspond to preferred sites of attack for an external proton. It is seen in Fig. 14.16 (see color section) that cholesterol protonates on the oxygen atom of the hydroxyl group, though a less reactive site is also observed in the vicinity of the double bond of the second cycle of the steroid.

We have found recently, however, that the MEP property is not sufficient to describe the reactivity of transition metal complexes, i.e. of molecules made of a metal atom, such as iron or chromium, coordinated to ligand atoms or groups (Weber *et al*, 1988a and 1989b). In this case indeed, a large charge transfer energy component has to be also accounted for, as the metal atom induces important orbital mixing or covalent effects between the inorganic or organometallic substrate and the reactant. This prompted us to develop a new theoretical model for the calculation of the interaction energy between an organometallic substrate S and an incoming reactant R characterized by its electrophilic (i.e. electron acceptor) or nucleophilic (i.e. electron donor) behavior (Weber *et al*, 1988a and 1989b). Within our model, the S-R interaction energy $E_{int}(\vec{r})$ used as a local reactivity index on the molecular surface of S is expressed as a sum of several components:

$$E_{int}(\vec{r}) = E_{es}(\vec{r}) + E_{ct}(\vec{r}) + E_{ex}(\vec{r}) \qquad (2)$$

where $\vec{r}$ specifies the position of the incoming reactant in the vicinity of the rigid substrate; E_{es}, E_{ct} and E_{ex} being electrostatic (i.e. the MEP when R is a proton), charge transfer and exchange energy components, respectively (Morokuma, 1977). In order to have at hand a fast procedure leading to the shortest possible response time on an interactive computer graphics system, approximate procedures have been developed to evaluate E_{es}, E_{ct} and E_{ex} and they have been reported in detail elsewhere (Weber *et al*, 1988a and 1989b). Suffices here to mention that the electrostatic component corresponds to the energy arising from the interaction of molecular, i.e. electronic and nuclear, charge distributions whereas the charge transfer term accounts for the energy lowering due to the delocalization of the electrons of one partner onto the other one as a result of their middle-range interaction. As mentioned earlier, this term is particularly important when S is a transition metal complex. In our model, both E_{es} and E_{ct} are evaluated within the fast and reliable extended Hückel (Hoffmann, 1963) molecular orbital framework. The exchange term, which describes the short-range repulsion due to overlap of both S and R electron distributions, is simply chosen to be zero outside the molecular surface of S and infinite on the surface itself (hard spheres approximation) which means that the minima of $E_{es} + E_{ct}$ on this surface are automatically taken as the most reactive sites (Goldblum and Pullman, 1978).

It is important to choose simple, though realistic, models for reactants since the computer time needed to evaluate E_{int} increases rapidly as a function of the complexity of R. In order to have E_{int} values depending only of the position of R with respect of S, and not on its orientation, two spherically symmetric reactants have been chosen: a naked proton as the model of an electrophile and a H^- hydride ion as the model of a nucleophile. This simple modeling is the same as that used successfully in previous theoretical investigations of reaction mechanisms in organic chemistry (Kahn *et al*,

1986). The model outlined here has been used for interpreting and predicting a large number of organometallic reactions and the results are in agreement with the experimental evidence as to the site and regioselectivity of the attack in more than 90% of the cases studied (Weber *et al*, 1983, 1988a, 1988b, 1989a and 1989b) . As an example, Fig. 14.17 (see color section) shows that the $(C_6H_6)V(CO)_4^+$ complex, made of a vanadium atom coordinated to a benzene ring and to four carbonyl ligands, should undergo a nucleophilic attack on the exo-face of the ring, i.e. on the face opposite to metal. This result is in perfect agreement with both the infrared and NMR experimental data reported by Calderazzo (1966), which confirms that simple theoretical models may be of great interest for a qualitative, and in some cases semi-quantitative, molecular graphics approach of chemical reactivity.

14.3 Conclusion

We have presented in this chapter several developments which have been recently achieved in molecular graphics in order to build, manipulate and animate models of chemical objects such as molecular structures and their related properties. In addition to their highly didactic and in some cases aesthetic aspects, these applications are of primary importance in molecular research, where it is essential to have at hand powerful computerized tools for the design of new compounds with specific properties. Through the extensive use of supercomputers and graphics workstations, it is indeed possible today to considerably help the design of new drugs, polymers or catalysts. The combination of theoretical calculations and graphics modelizations enables the chemist to guide experimentation in the rational prediction of novel species. In addition, the molecular graphics tool is invaluable in allowing the molecular scientist to interpret and understand important processes such as the binding of ligands to specific receptor sites on macromolecules or the adsorption of reactive species on surfaces, which could open new horizons in both basic and applied molecular research.

Acknowledgements

The authors are grateful to Professors C. Daul and E.P. Kündig for fruitful discussions, and to Professor J. Brickmann for providing a copy of the MOLCAD package. In addition, they have benefited from the assistance of their co-workers, Dr M.J. Field, Mr. P. Fluekiger and Mr. P.O. Regamey. This work is part of Project 20-25317-88 of the Swiss National Science Foundation.

214   *Jacques Weber and Pierre-Yves Morgantini*

References

Allen, F.H., Kennard, O., and Taylor, R., Acc. Chem. Res., 16, 146 (1983).

Brickmann, J., Chem. Des. Aut. News, 3(5), 7 (1988).

Burkert, U., and Allinger, N.L., Molecular Mechanics, American Chemical Society, Washington DC (1982).

Calderazzo, F., Inorg. Chem., 5, 429 (1966).

Catlow, C.R.A., in Computer-Aided Molecular Design (Ed. Richards, W.G.), IBC Technical Services, London (1989), p. 251.

Connolly, M.L., Science, 211, 709 (1983)

Dubois, J.E., in Computer Representation and Manipulation of Chemical Information (Eds. Wipke, W.T., Heller, S.R., Feldmann, R.J., and Hyde, E.), Wiley, New York (1974), p. 239.

Feldmann, R.J., Bing, D.H., Furie, B.C., and Furie, B., Proc. Natl. Acad. Sci. USA, 75, 5409 (1978).

Fischer, D., Ber. Deutsch. Chem. Ges., 27, 2985 (1894).

Frühbeis, H., Klein, R., and Wallmeier, H., Angew. Chem. Int. Ed. Engl., 26, 403 (1987).

Goldblum, A. and Pullman, B., Theoret. Chim. Acta, 47, 345 (1978).

Hoffmann, R., J. Chem. Phys., 39, 1397 (1963).

Jefford, C.W., Mareda, J., Combremont, J.J., and Weber, J., Chimia, 38, 354 (1984).

Kahn, S.D., Pau, C.F., Overman, L.E., Hehre, W.J., J. Amer. Chem. Soc., 108, 7381 (1986).

Karplus, M., and McCammon, J., Scien. Amer., 254, 30 (1986).

Lehn, J.M., Angew. Chem. Int. Ed. Engl., 27, 89 (1988).

Marshall, G.R., Ann. Rev. Pharmacol. Toxicol., 27, 193 (1987).

Max, N.L., J. Mol. Graphics, 2, 8 (1984).

McCammon, J.A., and Karplus, M., Proc. Natl. Acad. Sci. USA, 76, 3585 (1979).

Morokuma, K., Acc. Chem. Res., 10, 294 (1977).

Mueller, K., Ammann, H.J., Doran, D.M. Gerber, P.R., Gubernator, K., and Schrepfer, G., in Computer-Aided Molecular Design (Ed. Richards, W.G.), IBC Technical Services, London (1989), p. 119.

Quarendon, P., Naylor, C.B., and Richards, W.G., J. Mol. Graphics, 2, 4 (1984).

Rasmussen, K., Potential Energy Functions in Conformational Analysis, Lecture Notes in Chemistry, vol 37, Springer, Berlin (1985).

Rhodes, P., Chem. Britain, 21, 53 (1985).

Scrocco, E., and Tomasi, J., Adv. Quant. Chem., 11, 115 (1978).

Weber, J., Roch, M., Combremont, J.J., Vogel, P., and Carrupt, P.A., J. Mol. Struct. Theochem, 93, 189 (1983).

Weber, J., Mottier, D., Carrupt, P.A., and Vogel, P., J. Mol. Graphics, 5, 126 (1987).

Weber, J., Fluekiger, P., Morgantini, P.Y., Schaad, O., Goursot, A., and Daul, C., J. Comp. Aid. Mol. Des., 2, 235 (1988a).

Weber, J., Fluekiger, P., Ricca, A., and Morgantini, P.Y., in Visualisierungstechniken und Algorithmen (Ed. Barth, W.), Springer, Berlin (1988b), p. 17.

Weber, J., Morgantini, P.Y., Fluekiger, P., and Goursot A., in New Advances in Computer Graphics (Eds. Earnshaw, R.A., and Wyvill, B.), Springer, Berlin (1989a), p. 675.

Weber, J., Morgantini, P.Y., Leresche, J., and Daul, C., in Quantum Chemistry - Basic Aspects, Actual Trends (Ed. Carbo, R.), Elsevier, Amsterdam (1989b), p. 557.

Weber, J., Fluekiger, P., and Field, M.J., in Computer Animation '90 (Eds. Magnenat-Thalmann, N., and Thalmann, D.), Springer, Tokyo (1990), p. 21.

Weiner, P.K., Langridge, R., Blaney, J.M., Schaefer, R., and Kollman, P.K., Proc. Natl. Acad. Sci. USA, 79, 3754 (1982).

Wilson, S., Chemistry by Computer, Plenum, New York (1986).

Wright, J.D., Molecular Crystals, Cambridge University Press, Cambridge (1987).

Zauhar, R.J., and Morgan, R.S., J. Comput. Chem., 9, 171 (1988).

15

Graphics Visualization and Artificial Vision

Charles Baur
and
Simon Beer
Ecole Polytechnique Fédérale de Lausanne

This chapter addresses vision systems using graphic visualization and modelling for the representation of knowledge. In particular, the authors describe a vision system called MBVS (Model-Based Vision System) using 3D-CAD surface models for correlation. Starting with the a priori knowledge of a scene, the authors verify the presence and position of objects by comparison with hierarchical computer generated models. In this way is the simplification of the scene leading to a reduction in computing time for the analysis of the scene performed by another system working without models is achieved. In this way the image analysis of complex scenes is also improved. The models can be altered in complexity with the aid of an expert system; thus for every task a well-adapted model is in effect obtained, speeding up the entire correlation process.

15.1. Introduction

The field of visual sensors is expanding greatly. In spite of steadily increasing performances, industrial application has not remained abreast of expectations (Neumann). 95 % of industrial vision systems are confined to simple jobs like presence verification and quality control. A major reason for this is certainly the inability of most systems to interpret complex scenes, a task necessary for operations like pin-picking or guidance of a mobile robot.

The most difficult task for a system equipped with external sensors is certainly the perception and interpretation of its environment. To react "intelligently" to events and changes in its surroundings, it must first construct an internal model of the world. This model is composed of what the system knows about the world. Up to now, most of the proposed solutions have lacked an important feature : systems using artificial vision never encounter a completely unknown scene. They have always had a certain base of knowledge called "a priori" knowledge. For example, in the case of a mobile robot for the surveillance of a building, the walls and doors in the building have fixed locations and the type of the door will not change from one day to the next.

This means that such a robot needs to adjust a model to the perceived world, rather than generate an entire internal model itself. In fact a kind of world-model of things which are unlikely to change can be entered into the memory of the robot and compared to the information gathered by the sensors. This approach is commonly known as the top-down approach [Fig. 15.1]. It is evident that if enough information about the scene is provided, the top-down approach needs a minimum of computation. However, in the case of sparse information, this approach is very time consuming because a model must be provided for every possible orientation (Shirai 1987; Gruver et al. 1985). The reverse, the so-called bottom-up approach, first extracts all of the features such as edges and lines from the image without knowledge about the contents of the scene. This process (often called "low level vision") is known to be quite problematic.

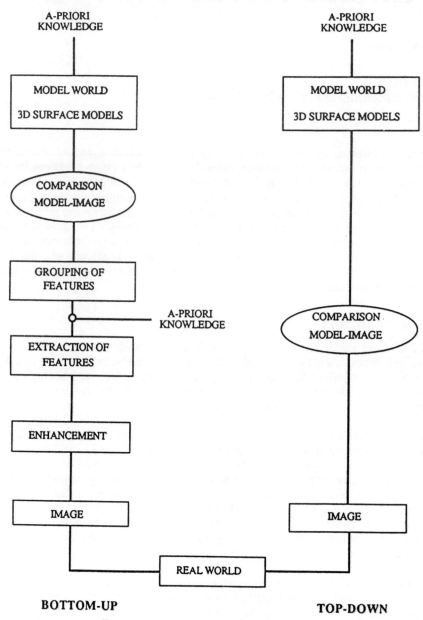

Figure 15.1. Top-down versus bottom-up

Grouping of the features, a process much more complicated than the former, takes place later. Only at this latter stage of the treatment does the a priori knowledge of the scene intervene. An ideal system combines the two approaches: eliminating all the objects known and/or defining regions of interest by a top-down analysis reduces the bottom-up system's task proportionally. Scenes with a high degree of uncertainty could also be processed by such a system.Several model-based vision-systems already exist. This chapter will mostly focus on systems using a graphic representation of knowledge more particularly the MBVS developed at the DMT (Département de Microtechnique) of the Swiss Federal Institute of Technology.

15.2. Related work

The way a-priori-knowledge is modeled provides a distinction between the different approaches. Most of the systems are application specific, few are related to mobile robots and so far none uses the full information of 3D surface models. This paper focuses on systems using geometric models.

A geometric model represents the complete shape and structure of an object. With this kind of modeling, one is viewpoint-independent and can easily deduce what features will be seen from any particular viewpoint and where they are expected to be. But one can also determine if a given image relationship is consistent with the model and what circumstances account for this specific relationship.

ACRONYM:
ACRONYM, developed by R.A. Brooks (1977) at Stanford University is certainly the most successful and best known model-based computer vision system. Interpretation of 2D images is effected by predicting geometric features provided by 3D models. These features are matched against the same features extracted from the image by low-level operators. The models are based on generalized cones, a volumetric element general enough to represent complex solids with only three parameters.

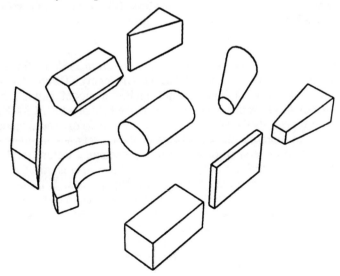

Figure 15.2. Some objects used by ACRONYM as primitives volumetric elements

All the objects are internally represented by a structure composed of different fields to which one can not only associate numbers but also variables. This way, it is possible to define classes of objects, which are composed of objects with the same form but different sizes. The objects themselves are embedded in an object graph. This graph defines the relations among the objects of a scene. The nodes of the graph represent the objects and the arcs the instances between the objects. By cutting this graph at a given level, ACRONYM defines different levels of models. These levels of models are used during the object recognition. The system refines the model only if the search for the coarser model has been successful. This technique eliminates too exhaustive a search for undesired objects.

Figure 15.3. Some levels of representation of an object

The main difference between ACRONYM and MBVS lies in surface models that make use of all the inherent information of these models (like texture,color, enclosed volume...). Results suggest that the surface information reduces data complexity and interpretation ambiguity. ACRONYM, on the other hand, reduces its volumetric models for the search to ribbons and ellipses. This is mainly due to its low-level feature extraction procedures, which can generate only this kind of feature for matching against the models.

PHI-1:
This model-based computer vision system was developed from 1985 to 1988 at the "Institut für Informatik und angewandte Mathematik" at the University of Bern (Gmür and Bünke 1988). It works by matching an image graph against a model graph. The 3D object representations used by this system are generated with the help of a PRIME-MEDUSA CAD-system. This system represents the volumetric objects by surfaces. These models, however, are not directly used to represent the knowledge of the objects but are transformed into a special representation in the course of an off-line procedure . Before implementing this procedure, the internal data structure of the Prime-Medusa system is transferred to a ASCII-file. The procedure itself can be divided into the following steps: 1. Determination of the surface normals to help further classification steps /2. Classification and labeling of the edges as convex or concave /3. Classification and labeling of the corners with the help of the edges /4. Extraction of the curved edges as polygons.

The combined geometric and feature-based description of the model resulting from this procedure contains two distinct graphs: the edge-corner-graph and the edge-surface-graph. The latter is not used during the hypothesis generation process, but serves for hidden-line removal in the verification procedure.

In order to obtain the image graph which is to be matched against the model graph during the recognition process, the greyscale image is processed in steps: 1. Applying the Gauss-Laplace-operator /2. Detection of the zero crossings in x- and y-direction /3. Connection of the edgepoints to lines /4. Segmentation of the curves, arcs and ellipses /5. Joining of the lines at the nodes and labeling of these nodes /6. Separation of overlapping objects and generation of the image graph, which itself contains several object graphs.

The recognition process is then effected in two steps. 1.Hypothesis generation working on the principle of partial graph matching is a bottom-up process. 2. Verification is done in a top-down manner. The models whose model graphs were found to have a partial match with the image graph during the first step, are now projected as best suits the common nodes of model and image. Applying this procedure to all possible models and orientations and evaluating a distance measure between the image graph and the projection graph of the model provides the best match between. During this process, full information about surface orientation is used to delete the hidden faces of the object. The different steps are represented in figures 4.i (Gmür and Bünke 1988). This two-stage process appears very efficient. Even missing nodes influence the overall performance only slightly. In contrast to MBVS the information about the surfaces is not fully used and the objects have no associated properties, like color, gloss etc. The off-line creation of the model graph seems to prevent really flexible use of the system. But the combination of bottom-up and top-down processes along with the use of a custom system for model generation make PHI-1 a very interesting approach.

IMAGINE I :

IMAGINE I (Fischer 1989), developed in 1986 by R. B. Fischer at the University of Edinburgh, uses object models composed of a set of surfaces geometrically related in three dimensions. Input data is a structure called a labeled, segmented surface image whose segmentation, depth information and labels are entered by a human operator. This structure is like Marr's 21/2 sketch (Marr 1982) and includes absolute depth and local surface orientation. Recognition is based on comparing observed and deduced properties (surfaces) with those of the model. Surfaces are described by their principal curvature with zero, one or two curvature axes. Surface segmentation is based on orientation, curvature magnitude and curvature direction discontinuities. Image segmentation such as the one done by IMAGINE I leads directly to partial or complete object surface segments. A surface completion process is implemented in order to attempt a reconstruction of the complete surface. After having evoked a model, another algorithm tries to achieve a model-to-data correspondence based on surfaces properties. This system is one of the first to use surfaces for object recognition and representation of knowledge. IMAGINE II, an upgraded version of IMAGINE I, is presently under development.

Other known systems do not generate geometric models at all and depend solely on descriptive models (Matsuyama and Hwang 1985). Thus, much of the valuable information present is lost. For example, all information concerning texture, color and shadows is no longer available. This is one reason why we believe that three-dimensional computer-generated surface models are worthy of investigation for the representation of a priori knowledge. Moreover, computer graphic systems are becoming more and more powerful while their price is dropping.

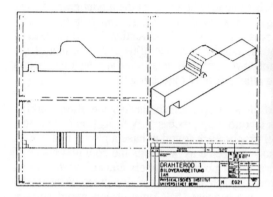

Fig.4.1 : CAD-model

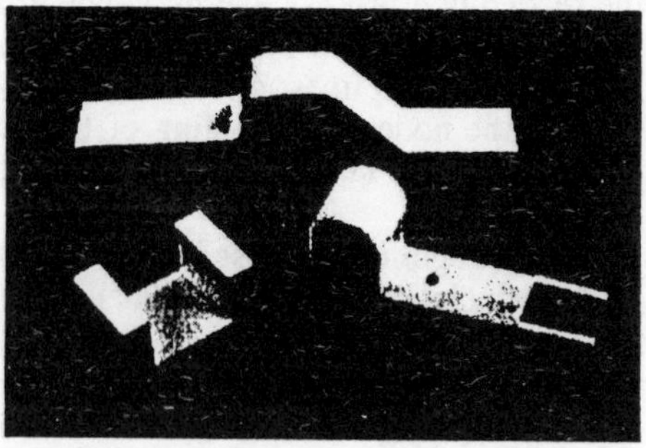

Fig.4.2 : Greylevel input image

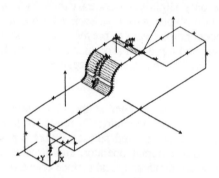

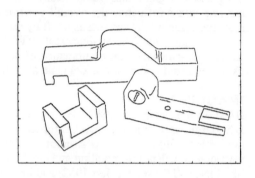

Fig.4.3 : Feature model

Fig.4.4 : Linked zerocrossings

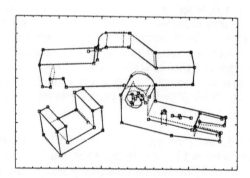

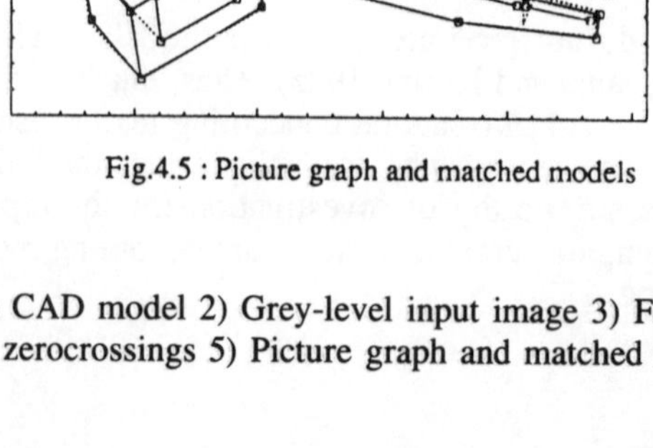

Fig.4.5 : Picture graph and matched models

Figure 15.4   1) CAD model 2) Grey-level input image 3) Feature model 4) Linked
zerocrossings 5) Picture graph and matched models

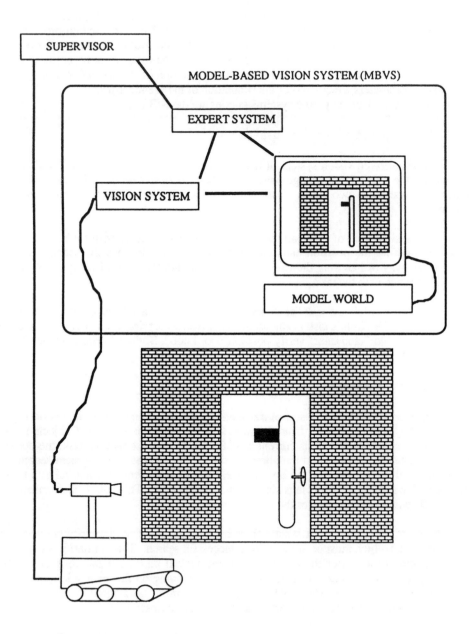

Figure 15.5. System overview

15.3. MBVS System survey

A detailed look at the overall system represented in Fig. 15.5, shows two parts not directly related to the MBVS presented in this paper: the mobile robot and the supervisor; they will not be described exhaustively here. Starting with the latter, the development of all of a supervisor's features is a task far to great to be incorporated in our research program and therefore will not be fully discussed in this chapter. The mobile robot itself does not influence the MBVS, but one important fact concerning the robot must be borne in mind. Thanks to its internal sensors, the position and the orientation of the camera are known at all times. These parameters are crucial for the simulation of the scene's aspect.
Now for a closer look at the partial systems making up the MBVS.

15.3.1 The supervisor (vision aspects)

MBVS is a task-driven vision system; it has the advantage that only the supervisor's global routines and the a priori knowledge have to be changed from task to task. Our laboratory applied this system to two fields : assembly and mobile robots. Only the latter will be discussed here.

From the vision point of view, all the tasks a mobile robot has to perform boil down to a single one : define the position and orientation of a 3D model in a 2D scene. Only the objective, the way of handling the result of correlation and the level of complexity (details modelled), change with the task.

Consider for example the task "open door", which could be a subtask of "go from room x to room y". For MBVS, this task is automatically driven by a priori knowledge, divided into "verify position door" and later "verify position door knob". Both tasks are governed by the same routine "verify".

15.3.2 Expert System

This part of the MBVS acts as a sort of translator between the supervisor, the vision system and the CAD system. It receives from the supervisor an object or a group of objects it must look for in the scene, as part of its current task. By taking into account the task, the position and orientation of the camera and the library of object models, the expert system searches in the hierarchical model tree for the appropriate level of representation and all simulation parameters like camera aperture, lens simulation, distances etc.. This data is then transmitted to the CAD station which produces the necessary image.

At the same time, the expert system fixes a minimum level of similarity between the object and model which must be attained for a successful search. This similarity is expressed by a correlation ratio. After the search the expert system checks, if the correlation ratio returned by the vision system is sufficiently high. Depending on this evaluation, the expert system determines its next steps, which may for example be adaptation of the model parameters or simply proceed if the search was considered successful.

The expert system we are working with is based on Nexpert from Neuron Data Inc.. In our case, this rule-based expert system shell runs on a Macintosh SE. The main advantage of this type of system is that it is easily extendable by simply adding a new rule or object. The rule format is a symbolic structure of the type:

if..... then .... and do .....

where *if* is followed by a set of conditions, *then* by a hypothesis or a goal which becomes true when conditions are met, *and do* by a set of actions to be undertaken as a result of a positive evaluation of the rule. Possible actions and the Rule Edition window are presented in Fig. 15.6 (NEXPERT OBJECT 1987). This same type of rule is also used in the Sigma system (Gmür and Bunke 1988).

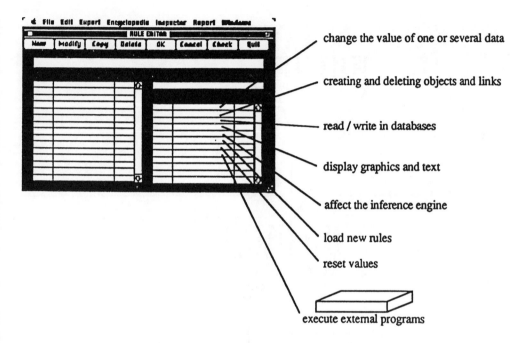

change the value of one or several data

creating and deleting objects and links

read / write in databases

display graphics and text

affect the inference engine

load new rules

reset values

execute external programs

Figure 15.6. Rule editor with types of actions

With the Nexpert Shell, the objects can be ordered in a hierarchical way. Considering that our models are also hierarchical, Nexpert proved to be a valuable instrument for our testbed. A part of our application's object tree is presented in Fig. 15.7.

One final important aspect of this expert system is its ability to treat the uncertainty. In this state often encountered in the field of image understanding, the system can work with incomplete data and still find a solution.

15.3.3 Artificial World

The function of this part of the MBVS is twofold:

- To preserve the last known or encountered state of the real world and thus function as a type of memory.
- To provide the images that will serve as models for the correlation.

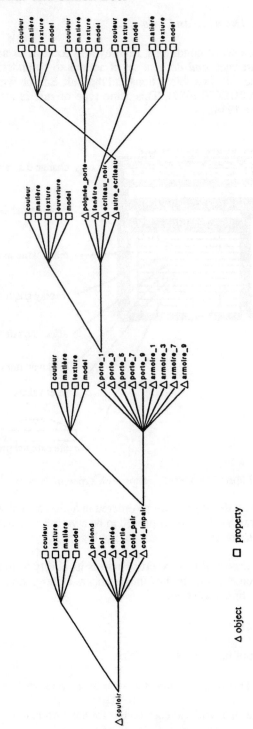

Figure 15.7. Hierarchical object database

This first aspect of the CAD database is very important for the MBVS. By preserving the last encountered state of the real world in the model space in most cases search time for the model in more complex scenes can be cut. Furthermore, a knowledge base for the supervisor to assist with decision making can be represented. Starting with a model of the last state implies correlation with a highly probable model, i.e only minor changes will have to be made to get a good match.

Objects in MBVS do not have just one model. Object models have different levels and may change with the task. This means that the models differ in resolution, number and representation of details, consideration of parameters like illumination, texture, gloss etc., but with only one geometric database. The higher the level, the more exact the representation; in other words, models can vary from a colored blob up to photorealistic image or other representation which is most suitable for the vision system [Fig. 15.8]. We believe that photorealistic models may not be the best models for the system and intend to investigate further in this direction to find the best adapted model for each task.

One of the main advantages of the use of CAD surface models is the possibility of rotating the models around each axis. Thus can a crucial problem of model-based vision systems be overcome : reducing the number of models needed when only sparse information about the scene and possible orientations of the object is available. With a CAD system the model can be rotated in a few milliseconds in any desired and realistic position. To speed up the whole process, each object can be associated with its possible degrees of freedom. For example, a door is typically an object with only one degree of freedom: rotation around the z-axis $(0,0,0,0,0,1)$. A telephone is encoded as an object with 4 degrees of freedom $(1,1,1,0,0,1)$. Certain elements of an object may inherit, among other properties, the degrees of freedom of their parent object or may have their own degrees of freedom. The body of the phone for example has the same degrees of freedom as the whole phone but the receiver has 6 degrees of freedom (see Fig. 15.9).

Thanks to the use of CAD models, it is also possible to take into account overlapping objects. Every possible realistic position of two objects can be represented. By realistic we mean in this case that with the aid of the expert system constraints for possible orientations and positions can be introduced. For example, it is normally impossible for a pyramid to rest on its tip. Thus a reduction of possible solutions is achieved, further speeding up the correlation process.

Correlation techniques have always encountered problems with non-rotationally symmetric objects as they are overly sensitive to changes in rotation and scale. The first of these problems can be reduced by the capability of rotating the object models. The latter can be reduced by a precise simulation of the lens characteristics, the camera position and its orientation. For most types of lenses, a simple linear function is sufficient for exact simulation.

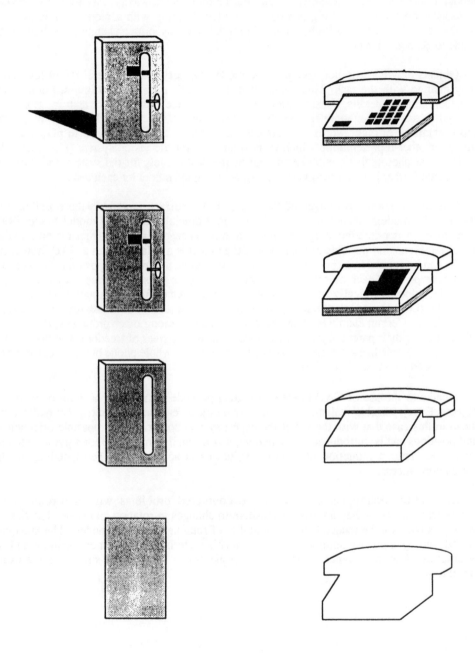

Figure 15.8. Possible levels of representation

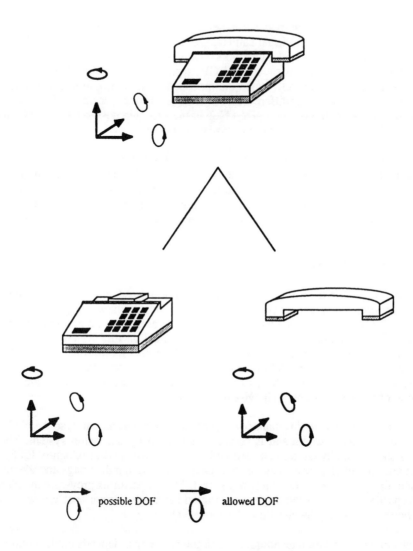

Figure 15.9. Inheritance of degrees of freedom for object and subobjects

Modern simulation techniques and computer graphics today allow the simulation and representation of all of the following aspects of an object in a reasonably short time:

1) Geometric data of the object
2) Position of the object (6 degrees of freedom in the most general case)
3) Position and orientation of the camera

4) Camera characteristics: focal distance, distortion
5) Surface qualities: texture, color, gloss
6) Illumination
7) Resolution

Using these parameters, our models can be perfectly adapted to the task at hand. The CAD system functions as a kind of virtual camera. Our system being greyscale-based, one might ask why one should bother simulating colors. We think that it may be very rewarding to use a color camera and treat each basic color (RGB) separately. The color planes of the graphic system can simulate the same effects on the models. It is furthermore possible to use filters on the camera to enhance color differences. For example, if one is searching for a green object in the scene one can first look for a green blob at the expected position and in its immediate vicinity and then refine the model only in case of ambiguity.

So far, most of the modeling work has been carried out at our laboratory on an APOLLO 580-T workstation. The modeling program is written in Fortran 77 using a library of 3D-Graphics routines. The library used is the Domain© 3D Graphics Metafile Resource (3D GMR), and it is running under P.H.I.G.S. (Programmers Hierarchical Interactive Graphics Standard). Its ability to give a hierarchical structure to geometric objects is ideally suited to our application.

15.3.4   Vision System

The vision system uses an algorithm based on normalized correlation. In the MBVS normalized correlation, a process also known as greyscale pattern matching, is performed on a Cognex EDS-3400 system. This technique allows the MBVS to be non reactive to global variations of illumination. CAD models are downloaded on this system via analog video signals (RS-170 compatible) and then correlated to the scene. It must be emphasized that, for the correlation, instead of a model of the entire scene, we use only that part that is the focus of interest. This imitates human behavior: a natural vision system only accurately analyzes the portion of the scene closest to its visual axis.

Downloading models in this manner is very fast (frame grabber rate), so the CAD station functions as a virtual camera and is considered as such by the vision system. The entire process is speeded up by the intrinsic properties of the above system (Cognex 1988). It uses a kind of artificial intelligence technique to scan quickly through the image and identify those areas most likely to contain the object of interest. Then it performs greyscale matching only in those specific places. Custom high speed digital hardware further accelerates processing. Typically recognition and localization take about 300 ms.

Thanks to this correlation technique, search time no longer depends on the complexity of the scene. Furthermore, it has already proved itself successful in other systems (Dickmanns and Graefe 1988; Dickmanns and Graefe 1988b). However, search time does depend on the complexity and size of the model. Both these parameters can be readily controlled by our system and therefore adapted and optimized to perform a given task.

Another great advantage of the correlation technique must be emphasized here. Correlation can be effected optically which means in real time. With the addressable holographic matrices now under development a truly fast system will soon be available with no change in either the overall approach or modelling techniques.

15.4. Application and results

After preliminary testings of the principle's validity (Baur and Beer 1989), we chose as an example a typical task for a mobile robot vision system: the search and determination of the position of a door in a corridor of our laboratory.

Using construction blueprints of the building, we first created a synthetic representation of the static elements in the corridor (wall, floor, wardrobes, lights etc.). Then we included the correct position of the model of a door. Our model of door has five possible levels of complexity. Defined as an object of one degree of freedom in the expert system, the door can be rotated around the z-axis in every possible position planned by the architect and permitted by the surrounding objects. The mobile object (one with two or more degrees of freedom) we chose was a table, which can be placed anywhere, thus simulating an obstacle. The camera and an additional light source can be positioned and oriented anywhere within the scene. Though only 256 colors are available and no transparency and shadows are possible, quite realistic images of the scene can be obtained [see Fig. 15.10, colour section)].

The small expert system installed for demonstration purposes is composed of some twenty rules and more than 30 objects. It is capable of verifying if the model of an object exists and then communicates directly with the vision system to initiate different tasks, then evaluate the correlation ratio returned. This installation shows that it is possible to correlate with the CAD-based surface models (Felder 1988). The door was successfully and easily found in the scene and correctly marked, as shown in Fig. 15.11 (see colour section).

Tests showed that with more details better correlation ratios can be obtained. However, the tests also demonstrated that the correlation ratio for a model with less details may still be high enough for certain tasks. This further strengthens the idea that task-adapted, hierarchical models are well suited for a vision system. It was further noticed that, in most cases, a coarse consideration of the illumination is sufficient for good correlation. However, the CAD-based surface models proved to be a very efficient way of storing a priori knowledge for a correlation-based vision system.

15.5. Developments in progress

We are currently starting to investigate the influence of photorealism on the correlation process. We have first evaluated different software products capable of modeling, rendering and animating 3D objects. We are now starting to work with the Explore software system from Thomson Digital Image (TDI). Thanks to the possibilities provided by this software, we were able to create and very efficiently adapt a 3D representation of a priori knowledge. Explore enables us to generate objects with a realism as pictured in Fig. 15.12 (see colour section). Our main task will be to determine the different levels of complexity for the models, then to optimize each one for the vision system's requirements. Furthermore, we are trying to incorporate other techniques, like structured light illumination to improve and accelerate scene analysis and interpretation. We will begin by researching the following two approaches to this technique by the line striping method :

1) Classical line striping: projection of a line onto a scene and analysis of its deflection with respect to geometric relations, thus providing values relative to the object's depth.
2) Correlation of the projected line with a model of the line created on the CAD system; this also involves inclusion of line models in the MBVS.

In addition, we want to profit from another development in progress in our laboratory: multiple line striping by holographic methods [publication scheduled for 1990]. We intend to use it to overcome some of the limitations of single line striping by projecting a structured light grid onto the scene. This way, depth information for an entire square region of the scene instead of simply one line can be obtained. This new technique accelerates the entire interpretation process considerably.

Acknowledgments

We would like to express our thanks to Professor C.W. Burckhardt, head of the Département de microtechnique of EPFL for his patient understanding of the purpose of our research.

References

Baur Ch, Beer S (1989) Bildanalyse mit Hilfe von CAD-Oberflächenmodellen. To be published in *Robotersysteme* 5, Springer-Verlag: Berlin

Bolles RC, Horaud P (1986) 3DPO: A Three Dimensional Part Orientation System. The *International Journal of Robotics Research*, Vol. 5, No. 3, pp. 3 -26, Fall (1986)

Brooks RA (1981) Symbolic Reasoning among 3D objects and 2D models. *Artificial Intelligence* 17, pp. 285 - 348.

Cognex (1988) Vision for Industry, Development Environment, Release 1.0

Dickmanns ED, Graefe V (1988) Dynamic Monocular Machine Vision. *Machine Vision and Applications* 1, pp 223 - 240

Dickmanns ED , Graefe V (1988b) *Applications of Dynamic Monocular Machine Vision.* Machine Vision and Applications 1, pp 241 - 261

Felder M (1988) *Systeme de vision à correlation normalisée utilisant des modèles 3D.* IMT-DME, EPFL

Fisher RB (1989) *From surfaces to objects*, John Wiley and Sons, 1989

Gmür E, ˚Bunke H (1988) PHI-1: Ein CAD-basiertes Roboter Sichtsystem. in *Proc. of Mustererkennung 1988*, 10. DAGM-Symposium, Zürich, Switzerland, pp. 240 - 247

Gruver WA, Thompson ChT, Chawla SD, Schmitt LA. (1985) *CAD Off -Line Programming for Robot Vision*. Robotics, vol. 1, pp. 77 - 87

Hwang V, Davis LS, Matsuyama T (1986) Hypothesis Integration in Image Understanding Systems. *Computer Vision, Graphics, and Image Processing* 36, pp. 321 - 371, 1986

Ikeuchi K (1987) Generating an Interpretation Tree from a CAD Model for 3D-Object Recognition in Bin-Picking Tasks. *International Journal of Computer Vision*, Vol. 1, Number 2, pp. 145 - 165

Marr D (1982) *Vision*, pubs : W. H. Freeman and Co, San Francisco

Matsuyama T and Hwang V (1985) SIGMA: A Framework for Image Understanding - Integration of Bottom-up and Top-down Analyses. *Proc. Int. Joint Conf. Artific. Intell.*, Los Angeles, USA, pp. 908 -915

Neumann B., *Wissensbasierte Konfigurierung von Bildverarbeitungs-systemen*, FBI-HH-M-147/87, Universität Hamburg

NEXPERT OBJECT (1987) Version 1.0, Neuron Data Inc., Palo Alto California, 1987

Shirai Y (1987)*Three-Dimensional Computer Vision*. Springer-Verlag: Berlin, Heidelberg, New York

16

Collaboration Between Computer Graphics and Computer Vision

André Gagalowicz
INRIA

16.1. Introduction

The scope of this chapter involves both applications related to computer vision such as vision for the robots of the next century and computer graphics such as audio-visuals or simulators of physical phenomena. It is a large field of applications where refined 3-D recognition and interpretation (not yet generally solved today) will be needed. It is related to medium term or long term applications. The problem that we intend to solve more precisely is to understand and reorganize a 3-D scene observed by two cameras in a stereo position, using an a priori given model of the scene. We have in mind the extension of this work to a temporal series of stereo images taken from various positions of the scene though we will not treat this problem now.

The originality of the method we present comes from the fact that we provide a feedback loop to do the image analysis. This feedback loop consists of the use of computer graphics to produce a synthetic stereo pair with the help of a model of the scene obtained by the analysis phase.

But, the problem that we want to solve is also to synthesize a very realistic 3D scene, realistic in the sense that if we see a given scene (or if a camera senses this scene) we will also produce a synthetic scene such that it will be impossible to discriminate visually between the natural and the synthetic images.

This is allowed with the use of an analysis phase of the scene where we will compute all the model parameters necessary to computer the image. Using a feedback loop between computer vision and computer graphics provides a way to solve simultaneously applications in two fields. Consider first the reasons which brought us to this approach.

If we observe Figure 16.1, we may note that image analysis uses as input an image space. Computer vision deals with the space of 3D scenes which is projected into the image space with the use of the focal plane of a camera (sensor). A proper acquisition technique transforms the analog image into a digital one. Digital image processing techniques transform digital images into themselves. These techniques are commonly used to enhance the quality of the processed images. So, their analysis provides models of the observed images or scenes.

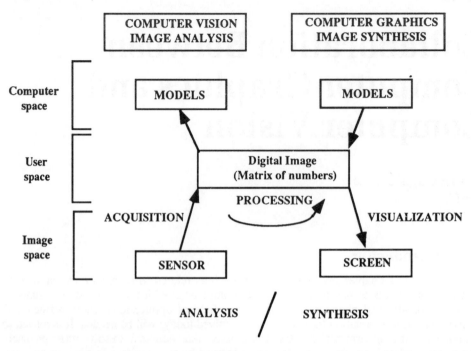

Figure 16.1. Computer Vision and Computer Graphics Duality

In computer graphics, the procedure is perfectly dual. It starts from scene models and by synthesis techniques produces digital images. Computer vision and computer graphics share in common the user space where computer programs do digital image processing. Visualization techniques give the possibility to present the digital images on a screen.

Though the two sciences, both very young, are completely dual, they did not "talk" to each other very much until nowadays. The two scientific communities still seem to ignore each other too much. Each science has its own purposes and applications and produces techniques adapted to them. So, these two sciences seem very distinct, but it is certain that each science could learn a lot from the other one. We all know the gains of the collaboration of distinct sciences to promote progress and innovation. Obviously, computer vision and computer graphics, being perfectly dual, the collaboration between these two sciences should be easier and more fruitful than between other domains. The experience accumulated by INRIA in texture modeling and fractals, where the collaboration between analysis and synthesis proved to be very innovative and fruitful, has shown us the power of this approach. We are convinced that this collaboration should be also very fruitful when applied to general 3D scenes.

Each science has its own strengths and weaknesses. The strength of computer vision is its ability to manipulate natural scenes, complex scenes. Unfortunately, the models created are generally very simple : this cell is, or is not, cancerous; such or such object does exist, or not, in the scene, etc.

Strength of computer graphics relies on its modeling capabilities. It is necessary to have **complete** models of the scene to produce images. Synthetic images became very spectacular which proves this modeling strength. The weakness of computer graphics is its heuristic nature which is the reason for the lack of realism in the gotten images. Even if realism is sought, we get only a pale illusion of reality. Computer graphics does not know how to copy natural scenes in general.

We do really believe that the collaboration between these two sciences should allow computer vision to converge to more complete models of the observed scenes, and computer graphics to reproduce more realistic natural scenes. We propose a procedure available for the simplified case of polyhedral scenes such as cities or offices well approximated by such surfaces. We believe that this procedure could be extended to more complex scenes.

The principle of the collaboration scheme proposed is presented in Figure 16.2.

16.1.1. Acquisition

Given a real 3D scene, we analyze this scene with two cameras, the optical axes that lie on a horizontal plane.We put both cameras in such a position that both optical axes form a horizontal plane and make an angle of 5 to 10 degrees, and such that the distance between the two cameras is small regarding the distance between the cameras and the objects of the scene.

The experimental conditions insure that all objects will be approximately projected in a cylindrical way on both images and that their projections will be very similar, in general.

16.1.2. Analysis

We want to study an analysis technique allowing us to produce a **synthesizable** model of the scene. This model should have two parts : a **geometric** part and a **photometric** one.

- The geometric part is itself subdivided in two types of information : the description of the **scene objects** based upon planar facets as we suppose a scene made of polyhedra, and the geometric description of the **light sources** (particular objects) including their localization and their spatial extent (particular facets). The classical computer vision result is the geometric description of the scene (first type of the geometrical part) which does not include light sources. This already clearly shows that the expected synthesizable model is more complex than what computer vision usually provides. Furthermore, most techniques only provide a set of 3D segments which makes it very difficult to use them to do a 3D interpretation of the scene (object description).

- The second part of the model, the photometric part, is, as with the first one, also subdivided in two types : the **reflectance properties of the objects** (or planar facets) and the **emittance properties of the light sources**. This second part is even less studied in classical computer vision. The aim to study such complex models may seem completely unrealistic, and it is a priori correct.

However, ignoring light sources and scene photometry causes a lack of information, useful or even necessary to solve analysis problems. We want to study what this information may bring to computer vision.

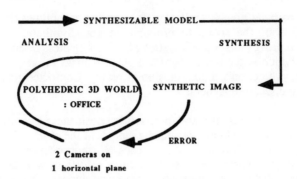

Figure 16.2. Analysis/synthesis collaboration scheme proposed

16.1.3. Synthesis

A model as previously described contains the necessary information to generate images. We know how to study and develop synthesis techniques using such types of models as input and produce images or stereo pairs of the scenes that they model.

16.1.4. Analysis/Synthesis collaboration

The idea of the analysis/synthesis collaboration is to compare images of the real scene coming from the retinas of the cameras and the synthetic stereo pair obtained by the synthesis program. This comparison effectively creates the feedback loop between computer vision and computer graphics. We foresee comparisons of two kinds : **qualitative** and **quantitative**.

- The qualitative comparison will be completed through the use of our brain. We shall observe both real and synthetic images and the errors observed will be interpreted by our brain to modify certain parts of the model and consequently modify adaptively the analysis and synthesis algorithms. We have applied this approach to textures (Gagalowicz 1987) and to fractals (Levy-Vehel and Gagalowicz 1988). Readers may consult the bibliography given in (Gagalowicz 1987) to verify the great number of innovative ideas this approach gave rise to these particular cases. We have noticed that it was always necessary to find an innovative procedure either in the analysis part, or the synthesis part (or even both) when we studied various models to converge to a correct solution. We are convinced that this analysis/synthesis loop should also bring much innovation when applied to this more general 3D problem.

- The second type of comparison is quantitative. All synthesizable models that we may produce will be always parameterized (object locations, facet locations, sizes, photometric parameters...). For a given model type, we may determine the best parameter values which minimize an error criterion (quadratic,...) between the natural and the synthetic images. This approach will be also used simultaneously with the qualitative approach.

We shall now describe the analysis strategy proposed to build a synthesizable model.

16.2. Analysis Technique Chosen

We shall only deal with a stereo pair, but the research we undertake is conceived for a future utilization for a set of stereo pairs varying over time, which could come from a robot moving in a room. The originality of the method results from the choice of **regional** approach versus the usual approaches where edge segments are manipulated. Readers interested in references about edge segment approaches may consult (Ayache 1988; Ayache et al. 1988). Stereo images are first segmented into regions. Matched regions produce 3-D facets which will be the basic primitives of our model. We have chosen this type of primitives because it corresponds better (than segments) to the need of geometric reasoning or 3D recognition. Furthermore, photometric properties of objects are intrinsically related to surfaces and not to edge segments, which reinforced this choice.

16.2.1. Image segmentation

The aim is to determine planar facets of the 3D scene. We have then chosen the option to directly look for projections of the facets in both stereo images, which means to look for regions having **similar homogeneous** properties in each image.

Image segmentation consists of determining a partitioning of the image into a set of connected regions such that each region satisfies some homogeneity constraints ("similar" pixel values). We hope that each homogeneous region obtained corresponds to the projection of a real 3D facet. Unfortunately, the homogeneity property which characterizes the projection of a particular 3D object is unknown. It is impossible to determine because of the complexity of lighting phenomena, of reflectance effects and of the lack of knowledge of the 3D geometry of the scene. The homogeneity property is clearly related to the reflectance values of the surface, a property that we should determine anyhow. Furthermore, this property differs from object to object. The crucial lack of knowledge forces people to design segmentation algorithms where only one general homogeneity property is used. Consequently, the results obtained are not satisfactory, in general.

These considerations have motivated our wish to put first all our efforts on a robust segmentation algorithm, robust in the sense that at least certain segmented regions will correspond to the projections of **true** 3D regions of the scene. We have shown in (Gagalowicz and Monga 1986) that it was possible to reach this goal while using a **hierarchical** region growing procedure.

The principle of the technique consists in choosing a criterion (related to the homogeneity property that we want to maintain) to merge adjacent regions. The criterion is computed for all of them and we merge sequentially the region pair corresponding to the best value of the criterion. Usual segmentation techniques merge region pairs as soon as the merge criterion is verified which produces scanning dependent results. A first segmentation is obtained through the use of a quadtree with the help of the MERGE procedure from Horowitz and Pavlidis (1974). We then build an adjacency graph based upon the adjacent regions obtained. The hierarchical procedure is then run recursively : at each step where two regions are merged, the adjacency graph and the region attributes used are updated. A menu of merging criteria is also available. Using less and less constraining criteria defines a second level of hierarchy where nested segmentations are produced. The robustness of the hierarchical algorithm may be evaluated on the results of Figure 16.3 (see color section). The two top images show the left and right images of a stereo pair obtained from an office of our laboratory (black and white images). The bottom images present the segmentation results of the corresponding images. Each detected region is visualized with a uniform color. It is easy to see that a certain number of segmented regions correspond effectively to projections of 3D surfaces. The segmentation is performed independently on both images.

The segmentation algorithms have been first developed for black and white images. We have then extended the procedure to color images with little increase in computing time and complexity (Vinet et al. 1989). We have also implemented a cooperative edge detection/segmentation approach (Vinet et al. 1989). For the latter approach, image edges, obtained from the Canny-Deriche algorithm (Deriche 1987), were used to control the segmentation : the segmentation obtained is such that all edges stay edges of the regions detected by the segmentation algorithm. We now have output information which **contains** the information given by conventional computer vision programs. It is richer !

The approach described above is a bottom-up approach : we start from pixels considered as individual regions and we merge regions while they are similar. This approach has the advantage of producing compact regions (without undesirable holes), but has the drawback that these regions have a tendency to overflow on neighboring **natural** zones. We may locally obtain either under or over-segmentations when compared to the ideal regions. The border of a segmented region may easily start to follow correctly the border of a physical zone of the image and then "dive" inside this zone or inside neighboring ones. So, we have also decided to test a **top-down** approach (Randriamasy et al. 1990). We first consider the total image as one region that we split recursively into two classes while using an automatic thresholding method coming from Kohler (1981). Each class contains a certain number of connected components themselves subdivided in the next step.

Figure 16.4 (see color section) shows the result of the top-down segmentation of the left image of Figure 16.3. The top left image is the original one visualized with pseudo-colors. The right one shows the result of the first division (first level of segmentation). The bottom left image shows the second one and the right one, the third level of segmentation. We get nested segmentations building a **segmentation tree**. The depth of the tree corresponds to the number of divisions performed.

This approach is interesting because it presents qualities and drawbacks different from the bottom-up approach. The top-down approach produces holes in regions : regions are not compact ; this effect is most of the time incorrect and undesirable. On the other hand, if we get a region which locally "seats" on the contour of a physical zone in the image, this will be the case over the entire border of the zone in general.

We now explore the possibility of creating a cooperative process between the top-down and the bottom-up approaches to take advantage of the complementary qualities of each of them, and get better results.

16.2.2. Region matching

In a second step we have to determine corresponding regions from the left and right segmentations: regions coming from the same 3D facet or region. As segmentation results are given (among others) by an adjacency graph, we have first proposed a solution (Cocquerez and Gagalowicz 1987) based upon sub-graph matching. This technique forces matches even if certain adjacency constraints are violated and, furthermore, possible matches are only one to one (a left region is matched with only one right region) which is often too constraining.

We have then proposed a more robust technique (Gagalowicz and Cocquerez 1987) which does not present the former drawbacks. The algorithm is itself decomposed into three phases:

16.2.2.1. Tree pruning algorithm

We use a tree pruning method which creates the most obvious matches in a very quick way. We use region attributes (surface, mean, variance, x-y coordinates of the inertia center,...) available from the segmentation results.

As matchable regions have to be similar because of the chosen experimental conditions, it is also the case of their attributes. We then choose a first attribute (size for example) and put the left and right regions in two heaps with decreasing values of size. For a given left region size, we allow a certain error on the right one, which dismisses most regions of the right heap. We then choose a second attribute and dismiss other regions from the remaining ones and so on.

The result of the tree pruning method is visualized in Figure 16.5 (see color section). Matched regions are shown with the some color on the left and right images. Non matched regions are painted in black (see Figure 16.4, for comparison).

16.2.2.2. "Single" regions determination

If we consider the camera geometry, it is easy to understand that the rightmost part of the field seen by the left camera is not seen by the right camera and conversely. It is then clear that some regions of the right border of the left image may have no equivalent on the right image and conversely. These border regions (and eventually others partly hidden) are called "single."

Practically, the two cameras are never perfectly horizontal and neither are the optical axes coplanar. Other single regions may then also appear on the top or bottom of each image depending on the real geometry of the cameras.

We show in (Gagalowicz and Cocquerez 1987) how to design simple rules to detect these single regions. Figure 16.6 (see color section) presents the result of the detection of single regions from the stereo pair of Figure 16.3. It is similar to Figure 16.5. The single regions are those which appear as new when compared to Figure 16.5.

16.2.2.3. Asymmetrical matching

Results of the tree pruning technique are used as an initialization to propagate new matching around already matched regions. The choice of this strategy is guided by the simple fact that, generally, if a region is next to a region of the left image and situated at a certain position with respect of this region, then its match should be explored among the regions next to the match of the region of the left image, and with a similar relative position with respect to it.

The principle of the method consists in following the border(s) of each region of a left and right match, called mother regions, and in building a heap obtained by the succession of regions met on the border of each of them (see (Gagalowicz and Cocquerez 1987) for more details).

The two heaps are then compared in parallel and recursively using a similarity criterion and allowing merges of p regions of the left image and merges of q regions of the right heap being matched. Generally, p and q are different integers which explains the adjective "asymmetrical" given to the method. In Figure 16.7 (see color section), we added regions matched by the asymmetrical matching technique to the results of Figure 16.6.

The results are still partial. We hope this method to become more efficient if we control in a more refined way the parallel displacement along the two heaps and the various configurations tested.

The remaining black regions correspond to the still not matched ones. It is due mainly to segmentation errors. Segmentation is the weakest and most basic part of the analysis method ; this explains the importance we give to this section.

Certain merges are done on the left image which are not produced by the segmentation on the right image and conversely. It is due to the fact that both cameras are not identical and don't see the scene from the same position. Consequently the images are different. Thresholds used to authorize merges are identical on the left and on the right images. This produces different results because of the differences between them.

It should be interesting to adapt thresholds locally on both images to improve the number and quality of matches. We propose a cooperative left/right segmentation to converge to this local adaptive approach.

16.2.3. Cooperative segmentation of stereo images

The segmentation and matching techniques described earlier produce a reasonable number of matches, but not enough to allow a global 3D interpretation of the scene. Persisting problems come from the fact that left and right segmentations may be sometimes locally very different because of the sensing geometry, the calibration of both cameras. A well contrasting zone of one image, where we can discriminate various objects, may appear bad contrasted in the other image, and no object may be noticed. The segmentations obtained are related to the choice of a threshold ε of a (uniformity) predicate ; this threshold is fixed by the user. It is arbitrary and produces very different results on the left and right images; these results are strongly related to ε and not always to the scene objects.

It should be interesting to adapt the segmentation so under-segmented regions (dark regions, or too few contrasted ones) could be split and over-segmented regions could be merged. The adaptive strategy we propose is an alternative possibly reducing all the deficiencies described above. The idea is to control the segmentation results of one image by the segmentation results of the other one and conversely. We propose the following heuristics :

A stereo pair is locally well segmented if a left/right match exists for the analyzed region

Segmentation and region matching are associated: segmentation is correct for an image if it is also correct for the other one and if matches were found. This heuristic permits us to envisage a segmentation process which no longer depends on the arbitrary choice of ε of the predicate; it is adaptive locally.

To avoid spliting and merging regions locally, we decided to build explicitly the segmentation tree of each image: for each region we create pointers to its father and to its sons.

If, to a clear or good contrasted left zone where we see different objects, corresponds a less contrasted right zone, instead of dividing the less constrated zone (which would have been merged incorrectly into a single region), we shall not try to split this region to seek new matches; we shall only have to analyze this region at a lower level of the segmentation tree (we have simply to explore its sons) so that no new image manipulation is needed.

Practically, we shall explore not only "horizontal" matching : matching corresponding to identical (last) levels of the segmentation tree of the left and right images (classical matching), but also crossed matching : matching between various levels of the left and right segmentation trees which correspond to various values of the ε parameter (see (Gagalowicz and Monga 1986) for more details).

The interest of this approach is that we define image zones where matchings are insured. Elsewhere, when no matching is available, we know that the segmentation is not valid, that we have to improve it. When compared to the non cooperative approach, we could improve the segmentation performance, but the increase of performance has not corresponded to our expectations. The reason is that, even if we work on several segmentation levels instead of one previously, these segmentation levels are still governed by a threshold which has no physical meaning; it is user defined. A physical region could be correctly formed **between** the segmentation levels given by the various thresholds : a physical zone may be over-segmented for a value ε_i and under-segmented for a new value ε_{i+1}.

To avoid this problem and to get a totally data driven segmentation tree we are now exploring matching between binary trees obtained by the description of all unions of two adjacent regions (until the tree root where the entire image is a single region). In this binary tree, all levels correspond to the value of the similarity function of the best adjacent regions. The binary tree is built **automatically**. This approach is also studied for the top-down segmentation where the segmentation is no longer binary, but is also automatically built, and completely data driven.

16.2.4. Stereovision: planar facets extraction

When two regions of a stereo pair are matched, we have to determine the 3D position of the corresponding planar facet (we suppose that it is a planar facet, though it may be wrong to do so!). To do the computation, we are supposed to know the geometry of both cameras (obtained through calibration of the cameras), and the positions of the two regions, which are the projections of the planar facet on the two focal planes of the cameras.

The classical stereo techniques match points or edge segments (middle points, or extremities of the edges are used) and do a simple triangulation similar to what a geometer does. As matchings are done pixel to pixel, errors on the 3D location of the original point may be important; furthermore, it is then needed to connect the obtained 3D points (or segments) to create higher order primitives for further recognition.

We have chosen a different option, a global technique, that we hope, should filter out, part of the 3D information errors existing with classical techniques. The idea is to consider regions globally : we do not want to look for individual pixel correspondences within both regions. We have produced two types of algorithms (see (Gagalowicz and Peyret 1990) for more details) :

16.2.4.1. Geometric algorithm

The first type uses only geometric properties of each region (surface, inertia center and axes...) and allows the computation of the 3D facet equation. The borders of the facet are given implicitly by the regions projected on both focal planes.

The principle of the procedure is the following : a plane $z = ax + by + c$ carrying the 3D facet is known through its two projections and is parameterized by (a, b, c). If we know the sensing geometry, we may, starting from the left region R_g, compute its 3D location based on a guess of the plane P and project again this facet on the right image. Let R_{dg} be this projection (see Figure 16.8).

We may compute the attributes (surface, inertia center, inertia central axes) of R_g. These three elements are concatenated in an attribute vector Car_g of size 6x1 (1 parameter for the surface, 2 for the inertia center and 3 for the axes). We show in (Gagalowicz and Peyret 1990) how to compute the attributes of R_{dg} denoted Car_{dg} using Car_g and the (a, b, c) parameters of the plane P. We may also compute the attributes of R_d denoted by Car_d.

The problem we solved is to determine the best values (a^*, b^*, c^*) which minimize the error between the two attribute vectors Car_d and Car_{dg} : $||Car_d - Car_{dg}||^2$. Regarding the type of perspective transform used, we may obtain either an explicit value of a^*, b^* and c^* or an implicit one, solved by a relaxation procedure. Practically, a simple gradient algorithm is sufficient as we start from a cylindrical solution (not far from the correct one) and as the quadratic error is differentiable with respect to the parameters.

We minimize $||Car_{dg} - Car_d||^2 + ||Car_{gd} - Car_g||^2$ while projecting the left region on the right plane, and the right region on the left plane to get a more symmetrical solution. This solution allowed us to verify that the method is precise for well segmented regions, but is sensitive to segmentation errors.

To alleviate this problem, we now study a solution where we minimize the non-overlap error between R_d and R_{dg}. This way, we do not consider parameters such as the inertia center and axes not consistent through perspective projection. We want to study the precision of both techniques and compare them to classical results as well.

Figure 16.9 (see color section) presents the image of a synthetic stereo pair built by the modeler ACTION 3D developed at INRIA. The interest of this scene is that the geometry of the scene is simple (polyhedral) and the geometry is a priori perfectly known. The computation of the 3D facets after the segmentation (see Figure 16.10, color section) and matching of both images is visualized on Figure 16.11 (see color section). We may see that the planar facets obtained seem well positioned in the 3D space. The use of synthetic images allows the computation of reconstruction errors and the comparison of various stereo algorithms.

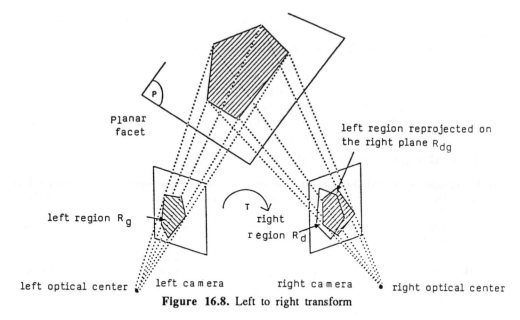

Figure 16.8. Left to right transform

16.2.4.2. Photometric Algorithm

The second type of algorithm only uses the intensity function of the region pixels. Till now, the problem we solved supposes that the planar facet is **lambertian** (ideally diffuse; i.e. diffuses the same energy in all directions). As in 16.2.4.1, the algorithm gives a solution (a*, b*, c*) for the plane carrying the 3D facet.

Given two matched regions (left and right) their intensity function is either constant or not. If it is constant and identical for both regions (with a given tolerance), then we may deduce that the facet is probably lambertian, as it produces the same energy from two different directions. This hypothesis is interesting as it allows us to model facet reflectance with only one parameter $\rho(x,y)$ (x,y : coordinates of a surface element on the surface). Furthermore, this reflectance is constant (=ρ) on the facet. If both functions are constant but different, the 3D facet is not lambertian. The photometric algorithm does not provide any solution here. In all other cases, the algorithm produces a solution (a*, b*, c*). This solution uses the property that the gradient of the intensity function of a region depends linearly of the normal to the plane P and of the gradient of the intensity function in the other region. The solution of P is then directly obtained from a simple linear system (Gagalowicz and Peyret 1990).

The interest of both methods is that they are new and may give some information about the reflectance of the facets.

For a given match, if both algorithms give about the same result, it will be a "proof" that the surface is lambertian. If they give different solutions, we could use this error in a feedback loop : if the facet is not lambertian, it is possible to model it simply by 2 or 3 parameters (the proportion of pure diffuse energy, of specular energy...) plus the geometry of the faces and light sources. We could find a choice of photometric parameters of the non lambertian facet which could minimize the error between both algorithms. The photometric algorithm will have to be adapted to the non lambertian case.

We intend to implement the geometric/photometric algorithms in the latter case, and compute a simple photometric model based upon the output of the geometric algorithm (that we suppose to give a correct solution), and upon the scene interpretation results of the next section which will produce the entire geometry of the scene (objects + light sources). This approach will be one of the tools that we intend to use during the feedback loop process.

16.2.4.3. Geometric interpretation of the scene

In this section, we intend to reconstruct the geometry of the observed scene from the a priori given 3D facets obtained by the stereo algorithms described above, and from an a priori given model of the entire scene. The problem we first intend to solve is restricted to the simplified case when we have at our disposal a global model of the scene where all objects are given by their absolute position in the scene : all objects are "clinched" in the 3D scene. This constraint will be eliminated later for objects such as chairs and desks, which will be modelled locally, and for which we will know their degrees of freedom in the scene. Till now, this solution is still unavailable as we do not have enough faces to recognize them individually.

The problem we want to solve is the following : given a polyhedral scene model and a set of faces given by the analysis of a stereo pair of a part of this scene, what is the displacement which will best match the latter faces with corresponding faces of the polyhedral scene model ? We are currently implementing a technique owed to Faugeras and Hebert (1983) which solves this problem.

The displacement simultaneously produces the scene interpretation in the two images. Each detected facet now corresponds to an object, and the scene becomes built up by closed objects.

To realize this experiment, we used ACTION 3D. This modeler allows to build a 3D scene **manually**, and to get all the geometry of the built scene. Furthermore, it also provides the position and extent of light sources and all photometric parameters used to produce the synthetic images (camera positions, focal lengths, focal axes orientation, light source emittance, object reflectance...).

With our procedure, it will be possible to determine the geometric description of the scene (light sources included) which will allow us to explore the photometric problems and to compare them with the "real" values provided by ACTION 3D.

Later, we intend to also explore the problem when light sources are not explicitly given in the 3D scene model; the new problem will be how to extract the geometry of the light sources from the scene geometry of the model and from both images.

16.2.5. Extraction of the scene photometry

We use as input the 3D geometric interpretation of the scene and the two images of the stereo pair. As output, we want to determine the emittance values of the various light sources and the reflectance functions of all 3D facets (of the various polyhedra). We will suppose that, to a given polyhedron, corresponds a single reflectance function (each object is supposed being formed of only one material !).

16.2.5.1. Lambertian model

The first photometric solution we propose corresponds to the lambertian case (all objects are lambertian). This study starts in our laboratory.

We shall suppose to begin with that the facet radiosity is only resulting from the direct lighting from light sources (+ ambient light, if necessary). We shall neglect energy transfers between lambertian facets, but it will be possible to take in account occlusion.

We shall use a pointwise approximation of the extent of the light sources. When a facet is lambertian and not textured ($\rho(x,y) = $ constant , $\forall$ $(x,y) \in$ planar facet), then its intensity seems constant when it is projected on any image plane(first order approximation). We will consider, first, non textured surfaces (which is a simplification) for which the reflectance may be approximated by a constant. This is often the case for objects in an office!

We may show here that it is possible to produce a compatible solution with one ρ value per facet and one emittance value E_i per light source, knowing the geometry of the scene and the mean of the intensity function of each region of both images. The solution may be refined if we have available several facets from the same object because this brings new constraints to this highly under-constrained problem. The lambertian model projects **constant** intensity functions on each pair of matched regions of the image planes; this constant is the mean of the intensity functions of the input images (computed on the union of the two regions). If this error is too great, it may be due to two reasons: presence of texture which can be detected on the image planes themselves (spectral signature is high) or shading effects resulting from the mutual influence of lambertian facets (if they are lambertian!).

The **radiosity solution** (Cohen and Greenberg 1985) which will be briefly described in the section 16.3.2.3 gives a solution to the problem of the mutual influence of lambertian facets. It allows us to produce intensity gradients (image getting darker to the volume corners...). The radiosity solution is the second step that we foresee to implement in the feedback loop.

16.2.5.2. More realistic model

If the radiosity technique still produces too many errors, we want to move to a more refined reflectance model incorporating a specular component. We first intend to use a simple Horn model having 3 parameters modelling two additive components : a pure specular component + a pure diffuse (lambertian) component.

This refinement may be completed from two sides : the use of both geometric and photometric stereo algorithms and the global analysis/synthesis feedback loop.

For the near future, we do not intend to study textured regions although some performing tools (Gagalowicz 1987) are already available to move in this direction.

16.3. Synthesis

Synthesis algorithms available in the literature (Foley and Van Dam 1984; Péroche et al. 1988) allow us to use, as input, models of the type produced by the analysis task.

16.3.1. Scene and animation modeler

We developed a new modeler (ACTION 3D) on an IRIS workstation and with the collaboration of SOGITEC. This modeler allows the interactive construction of synthetic scenes and their animation. The particularity of this modeler is its ergonomics. It is very interesting to use in an analysis/synthesis feedback loop, as we produce scenes where the entire geometry is available. We may use it to calibrate all analysis algorithms (segmentation, matching and 3D reconstruction) as we know a priori the ideal data to find.

16.3.2. Rendering algorithms

16.3.2.1. Scan-line algorithm

In the framework of the INRIA/SOGITEC collaboration, we have a scan-line rendering algorithm allowing the visualization of the models produced. This algorithm produces already very realistic images and it is suitable for a feedback loop as it is computationally very efficient. This algorithm may render scenes produced by ACTION 3D, and will be used to render scenes obtained from the analysis. The richness of the input photometric models (compared to the first ones we studied in section 16.2.6) should enhance the realism of the synthesized images.

16.3.2.2. Ray tracing algorithm

The ray tracing technique is an attempt to follow the trajectories of light rays. We consider the optical center of the observer and trace a ray starting from the optical center and hitting each screen pixel. This ray shall hit an object. From this impact point we draw two new rays: a specular one, reflected on the object surface according to the mirror law, and a refracted one according to Descartes' law. Each ray is recursively divided in two when it hits a new surface. After a certain depth of the binary tree obtained, we stop the search and trace the rays backwards using all object colors to produce the final pixel color. Because of its principle, this algorithm is very efficient mainly for specular and translucent objects even if some extensions were brought to the technique.

A new ray tracing algorithm using models built by ACTION 3D has been designed in the laboratory (Devillers 1988). This algorithm needs heavier computations than the scan-line algorithm but brings new realism to the images. Shadows are naturally obtained which is not the case for the scan-line procedure. Figure 16.12 (see color section) is an example of a ray traced image. It was designed at IRISA by Arnaldi et al. (1987). Similar results may be obtained by our own procedure.

16.3.2.3. Radiosity technique

This technique has been introduced at Cornell University by Cohen and Greenberg (1985) and is dedicated to lambertian surfaces. It considers globally all rays leaving a lambertian facet, which reach another one, without trying to differentiate the rays. A simple energy balance equation is given for each facet :

emitted energy + received energy = sent energy

The solution of this set of equations gives the radiosity of each facet (its color).

16.3.2.4. Ray tracing/radiosity mixed approach

Ray tracing has been already widely studied in the literature and efficient algorithms are available. On a sequential machine like a VAX, it needs hours to generate a ray traced image but ATT has proposed a massively parallel machine in 1987 which produces these images almost in real time.

The radiosity technique is new (dated 1984 roughly). The first algorithm required days of computation on a sequential machine. The numerical and hardware innovations of the last 4 years allow us to foresee a real time machine for the near future : Pr. Greenberg's team has already produced a seven images per second version in 1989.

The next step to reach complete realism is to combine ray tracing and radiosity procedures to deal with more general surfaces both specular and diffuse. Several solutions have been already proposed recently (Sillion and Puech 1989). We are exploring our own solution which will be the final algorithm that we intend to implement in the feedback loop. But, we need time, numerical efficiency and hardware progress before we can continue.

16.4. Conclusion

We have proposed a new approach to eventually solve computer vision problems and simulations of natural scenes. The solution to these problems seems visible even if it is still very far away. We do think that the analysis/synthesis approach owns the necessary power to solve very hard problems related to 3-D scene interpretation which may concern robotics of the next century. But it is also a path to synthesize very realistic scenes which can offer fantastic possibilities to mix dreams and reality.

References

Arnaldi B, Priol T, and Bouatouch K (1987) A New Space Division Method for Ray Tracing CSG Modelled Scenes, The Visual Computer, vol. 3, N. 2, pp. 98-108.

Ayache N (1988) Construction et Fusion de Représentations Visuelles Tridimensionnelles ; Applications à la Robotique Mobile; thèse d'état, Université de Paris-Sud, Orsay.

Ayache N, Faugeras OD, Lustman F and Zhang Z (1988) Visual Navigation of a Mobile Robot, in *IEEE Workshop on Intelligent Robots and Systems (IROS'88)*, Tokyo.

Cocquerez JP, Gagalowicz A (1987) Mise en Correspondance de Régions dans une Paire d'Images Stéréo, Proceedings du Congrès MARI 1987, La Vilette, Paris.

Cohen MF, Greenberg DP (1985) The hemi-cube, a radiosity solution for complex environments, Proc. Siggraph '85, Computer Graphics, Vol. 19, N.3, pp.31-39.

Deriche R (1987) Using Canny's Criterion to Derive a Recursively Implemented Optimal Edge Detector, International Journal of Computer Vision, pp. 167-187.

Devillers O (1988) Méthodes d'Optimisation du Tracé de Rayons, Thèse de Doctorat en Sciences, University of Paris Sud, Orsay, France.

Faugeras OD, M. Hebert M (1983) The representation, recognition and positioning of 3D shapes from range data, Proc. of the 8th International Conference On Artificial Intelligence, pp. 996-1002, Karlsruhe, BRD, August 1983.

Foley J, Van Dam A (1984) Fundamentals of Interactive Computer Graphics, Addison Wesley.

Gagalowicz A , Peyret M (1990) Reconstruction 3D basée sur l'Analyse en Régions d'un Couple Stéréo (in preparation).

Gagalowicz A, Cocquerez JP (1987) Un Nouvel Algorithme de Mise en Correspondance et Régions dans des Paires d'Images Stéréo, Proceedings du 6ème Congrès AFCET sur la Reconnaissance des Formes et l'Intelligence Artificielle, Antibes.

Gagalowicz A, Monga O (1986) A new approach for image segmentation, Proc. 8th International Conference On Pattern Recognition, Paris.

Gagalowicz A, Texture Modelling Application (1987), *The Visual Computer*, Vol.3, pp. 186-200.

Horowitz JL and Pavlidis T (1974) Picture Segmentation by a Direct Split and Merge Procedure, *Proc. 2nd. Intern. Joint Conference on Pattern Recognition, ICPR 74*, pp.424-433.

Kohler R (1981) A segmentation System based on Thresholding, Computer Graphics and Image Processing, vol. 15, pp. 319-338.

Levy-Vehel J and A. Gagalowicz A (1988) Fractal Approximation of 2-D Objects, *Proc. Eurographics '88*, pp. 297-312.

Péroche B, Argence J, Ghazanfarpour D, Michelucci D (1988) La Synthèse d'Images, Hermès Eds, pp. 199-238, Paris, France.

Randriamasy S, Kaswand T, Gagalowicz A (1990) Segmentation descendante par seuillage recursif (in preparation).

Sillion F, Puech C (1989) A General Two-pass Method Integrating Specular and Diffuse Reflection, Proceedings of SIGGRAPH 89 Conference, vol. 23, N. 3, pp. 335-344.

Vinet L, Sander PT, Cohen L and Gagalowicz A (1989) Cooperative Segmentation and stereo Matching, in *Topical Meeting on Image Understanding and Machine Vision*, North Falmouth, Massachussetts.

17

Graphics Simulation in Robotics

N.F. Stewart*
Université de Montréal

This chapter has two sections. In the first we shall give an introduction, and describe the general goals of robotic simulation. The second is divided into several subsections, describing various problems and methods associated with the simulation of robots, with emphasis on graphical simulation; these include the kinematics problem, the problem of collision detection, and the simulation of sensors.

17.1. Introduction

An important application of graphical simulation is the simulation of robots. Robotic development systems including graphical simulation are of considerable interest, for several reasons. For one thing, such systems permit evaluation of the robot before it (and/or related equipment) is purchased, and perhaps even before it (and/or related equipment) is built. This is the idea of "electronic prototyping" {see Chapter 4}. Secondly, they permit the programming of the robot, once it is built, away from the dangerous environment in which it must operate. Thirdly, once the robot is installed, the risk of damage (to people and to equipment), caused by inevitable programming errors, is reduced if it is possible to observe the execution of the task on a graphic screen before it is executed in the real world. This is especially important in the context of training people to program robots. Finally, such systems reduce the amount of time that an assembly line must be shut down when a robotic task is reprogrammed.

Many aspects of Solid Modelling {see Chapter 4} are relevant to robotic simulation, because we want to model both the robot and its surrounding environment as three dimensional solids. However, there are two general subfields involved here: *work cell design* and *interactive simulation* of objects in motion in robotic environments. (There is some overlap between the two, since the former (Stauffer 1984) may include kinematic modelling of the cell.) The complexity of the latter problem is greater since it may involve the simulation of complete robot programs, and the importance attached to various solid modelling problems is different: of the operations listed in [see Sect. 4.3), for example, the problems of interference checking, calculation of distance between objects, and of course, fast graphical display of moving objects, take on special importance. The emphasis is on *animation* (Magnenat-Thalmann and Thalmann 1985), rather than the more passive aspects of work cell design (such as whether a certain position can conveniently be reached by the

* This research was supported in part by a grant from the Natural Sciences and Engineering Research Council of Canada

robot), and on tools to facilitate interactive use of the simulation system. Real-time animation is not essential: slow motion display of a robot's motion may in fact be more useful, provided of course that is is not too slow. In many applications, simple kinematic animation (see Subsection 17.2.3, below) is sufficient.

A summary outline of some of the commercial robotic simulation systems available in 1984 was given in (Stauffer 1984), and there exists an excellent (and more recent) survey of the more general area of solid modelling (Johnson 1986). (One system described in (Johnson 1986) that has had considerable application in Robotics is the CATIA system.) Many simulators described in (Stauffer 1984) use wire-frame models: for example, MCAUTO, ROBOT-SIM (see also (Novak 1984)) and ROBOCAM (see also (Craig 1985)); some, however, like IGRIP (Mahajan and Walter 1986) use surface representations and implement hidden surface displays.

More theoretical areas of research, closely related to the problem of robotic simulation, include the study of higher level reasoning about shape and kinematic function (Joskowicz 1988), and the problem of motion planning. Good introductions to the latter area are (Yap 1987, Schwartz et al. 1987, Sharir 1989). Even an apparently simple problem like moving a circular robot in the plane, amongst obstacles with boundaries that are segments of lines or circles, turns out to involve a large amount of non-trivial mathematics. The problem of planning the motion of an articulated robot arm in E^3 is a very difficult mathematical problem.

17.2. Methods Relevant to Graphical Simulation in Robotics

17.2.1. Solid Modelling

We begin by emphasizing that many of the problems and associated solutions of solid modelling {see Chapter 4} will be relevant in the context of workcell design and robotic simulation; the criteria (and the problems viewed as critical) will vary, depending on the application.

For example, much of the emphasis in solid modelling for mechanical design {see Chapter 4} is on the ability to model objects of very complicated shape. However, in the context of a robotic simulation system it may be perfectly adequate to restrict the representation method to a CSG system with convex polyhedral primitive objects, and only one Boolean operation, namely set union (Hurteau and Stewart 1990) (see also (Bobrow 1989)). Fine detail is not necessary in order to evaluate different strategies for moving an object around a work space, and the simplified system just described is certainly sufficient for the geometric description of the robot itself. (Of course, these comments would not be valid in the context of fine assembly using robots.

On the other hand, other problems now become critical. For example, it is necessary to specify mechanisms and linkages; it is necessary that operations be carried out at speeds which permit interactive use of the system (which explains why we may be prepared to accept theoretically unsatisfactory {see Chapter 4} representations like wire-frame models); and operations like intersection (interference checking) become very important. These and other questions will be discussed in subsequent subsections.

17.2.2. Computer Animation

As indicated in the introduction, the techniques of Computer Animation (Magnenat-Thalmann and Thalmann 1985; Wilhelms 1987a; Wilhelms 1987b) are highly relevant to some aspects of Robotic Simulation. Most of these techniques will not be discussed here, but for completeness, they must be mentioned. More details may be also found in Chapter 9.

One important comment is that in general the animation problems in Robotic Simulation are relatively easy (compared for example to the problems of animation of the human body, natural phenomena like fire and smoke, and so on (Magnenat-Thalmann and Thalmann 1985)). Usually, motion specification (Magnenat-Thalmann and Thalmann 1985, p. 77) can be restricted to rigid motions and virtual camera motions, neglecting things like alterations of shape, size or colour of an object. For display, as mentioned above, a wire-frame presentation may well be sufficient, and if a boundary representation is used, display techniques can be restricted to hidden surface calculations, neglecting things like reflectance, shading, transparency, texture, shadows and anti-aliasing. To introduce these may be desirable (see for example (Schacter 1981), where texture is used in the closely related area of flight simulation), but in making the trade-off with cost and display speed, it may well be decided that they are less important. Similarly, although simulation of deformable and flexible objects is desirable (Gourret et al. 1989, Moore and Wilhelms 1988), it is not yet possible to do this in complete generality in real-time or near real-time.

17.2.3. Kinematic and Dynamic Analysis

One of the problems mentioned above was the specification of mechanisms and linkages. This is a problem that attracted the attention of engineers long before the advent of robotics.

A description of a typical procedure in the layout of a mechanism is given in (Johnson 1986, p. II-12): "... the designer may specify a motion of the driver part; a kinematic analysis is performed, and the resulting motion of all of the driven parts is computed. This kinematic analysis can be performed as an interactive function during layout...." The next step is the calculation of loads: "...For the robot example, the motion of the robot is computed by simulating its response to the robot's program. Mass properties are automatically computed from the solid models of each component. These properties are then used to compute loads on the robot's joints at each point in time in the simulation....Mass property calculation programs can be used to compute some loads, while friction loads (which are a function of the material properties and clearances of mating faces in the robot's joints) will require additional information to be associated with the solid model...." However, as stated in the first paragraph of the introduction, simulation of a robot may be of interest not only at the design stage, but at many later stages too.

A standard system for defining the relationship between the links in a mechanism is the Denavit-Hartenberg system {see Chapter 9 or Paul 1981, Ch. 1; Fu et al. 1987, p. 36) based on homogeneous coordinates. Given a description of the robot arm, including the coordinates of the joints, the problem of finding the coordinates and the orientation of the end effector of the robot is the *direct kinematics problem*. The inverse problem of finding the joint coordinates, given the position and orientation of the end effector, is the *inverse kinematics problem*. More complicated is the problem of robot *dynamics*, which takes into account forces and torques, and velocities and accelerations, and dynamic *control* of the robot.

References for the mathematics involved in these problems are (Paul 1981, Fu et al. 1987, Armstrong et al. 1987). A description of one approach to dynamic simulation, whose principles could be applied in a robotics context, and an entry to the associated literature, appears in (Moore and Wilhelms 1988).

Theoretical results on the complexity of rigorously establishing the feasibility of various motions are given or referred to in (Schwartz et al. 1987).

17.2.4. Collision Detection

One important characteristic of a robotic simulation system is that it should permit detection of collisions, and this capability is either existent or aimed for in commercial simulation systems (Stauffer 1984). Mathematically, this is the problem of the computation of the intersection between two objects; quite a lot is known about the problem in the two dimensional case, and somewhat less in the three dimensional case (Sharir 1989, Sect. 5; Levin and Sharir 1987, Sect. 7; Preparata and Shamos 1985, Ch. 7). The problem of finding the intersection of two *convex* polyhedral subsets of E^n is the problem of finding a feasible solution to a linear programming problem (Luenberger 1973). However, for small n, this approach may not be optimal, and if the objects are not convex, the best approach to the problem of finding convex triangulations (and even the definition of "best") is not necessarily obvious, even for n = 2 (Preparata and Shamos 1985). (We may note here that results for E^2 can be useful in motion planning and collision detection in robotics. Since vertical assembly is an important subclass of assembly problems, robots such as those in the SCARA class can be usefully modelled as two dimensional objects moving in the plane: see Figure 17.1. Furthermore, if we are dealing with mobile robots moving on the ground, rather than with robot arms, two dimensional models are appropriate. On the other hand, it may be of interest to introduce time as a fourth dimension, and compute intersections in E^4.)

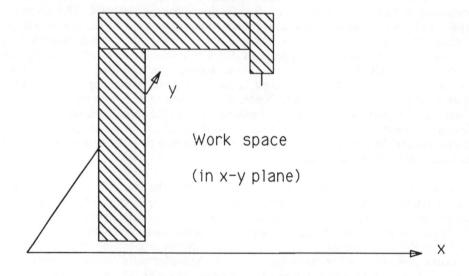

Figure 17.1. SCARA type robot modelled in the plane

The problem of collision detection may be considered in either the static or dynamic context (Moore and Wilhelms 1988,Boyse79) (*cf* remark above on representations in E^4. Of course, static methods are normally applied repeatedly, thus providing a method which may be viewed as a *discrete* dynamic interference method. If the time discretization is too coarse, however, there is a risk of missing a collision because one object has moved through another in a single time step [p. 292](Moore and Wilhelms 1988). Collision detection viewed as an intersection calculation for CSG-trees {See Chapter 4) (based on boundary evaluation) is discussed in (Tilove 1984), and (based on bintree conversion) in (Samet and Tamminen 1985).

One important technique in reducing the cost of collision detection algorithms is the use of bounding regions (Cameron 1984) and hierarchic indices (Mantyla 1988, Sec. 18.2; Samet 1990a, p. 355-359).

Collision detection algorithms are also of interest in Computer Graphics applications other than robotic simulation, of course (one example is the intersection between a line segment and an object, which is of interest in ray tracing; video game design (Uchiki et al. 1983) is another), as well as in other areas (for example, VLSI design verification (Samet 1990a, Ch. 3)). Intersection and fit in the context of Solid Modelling are discussed in (Johnson 1986, p. II-13). The intersection problem is also of interest for the problem of sensor simulation in robotics, which is discussed in Subsection 17.2.6, below.

17.2.5. Distance between objects

Related to the problem of collision *detection*, discussed in Subsection 17.2.4, is the problem of collision *prevention*, by which we mean the capability to alert the user in advance to the possibility of a collision between two objects in the robotic environment (Stauffer 1984, Hurteau and Stewart 1990, Bobrow 1989, Red 1983). The mathematical problem here is essentially one of computing the distance between two objects represented in the modelled robotic environment. Whether a wire-frame or hidden surface model is used, if the environment is at all complex it will be impossible for the user to monitor all parts of the environment: wire-frame displays quickly become so cluttered that it is difficult to distinguish objects well enough to estimate distance, and a hidden surface model by definition may hide certain (parts of) objects. We are thus led to the idea of a system which might, for example, display in a different colour that part of the robot which is closest to other objects, or alternatively, effect a visual classification of objects (such as red for objects within a certain prescribed tolerance, and other colours for more distant objects).

The problem of computing the distance between two objects is a topic of major interest in motion planning (Yap 1987, Schwartz et al. 1987, Cameron 1984), which was mentioned above. Straightforward branch and bound algorithms for computing distance between unions of polyhedral objects (Hurteau and Stewart 1990) can lead to surprisingly difficult mathematical problems (Andersson et al. 1987,Andersson and Stewart 1988). An algorithm for computing the distance between two objects represented by non-aligned octrees (Samet 1990b) is given in (Major et al. 1990).

17.2.6.   Sensor Simulation

Another desirable characteristic of a robotic simulation system is the ability to simulate sensors (Stauffer 1984, Pai and Leu 1986, SIEL 1987).

In the simplest case, such as that of a simple photo-detector, sensor simulation reduces to the detection of a collision between a pseudo-object, corresponding to the sensing envelope, and a real object. The mathematical problem is therefore formally the same as the one discussed above, in Subsection 17.2.4. Infrared detectors have been modelled in this way (Segura and Schrive 1985, Andre 1983), for example as part of the ongoing development of the CATIA system. In other cases, however, the problem is much more difficult. For example, it would be desirable to be able to simulate the operation of a vision system; some efforts have been made in this direction, in the area of synthetic vision using ray tracing or other methods (Raczkowsky and Mittenbuehler 1989), but this approach is much too slow for the simulation of industrial processes in interactive mode.

Other sensors, between these two extremes, admit a range of possible models. For example, inductive and acoustic sensors can be modelled very approximately using the range information provided by the manufacturer, by means of a more detailed model like the ones described above for infrared sensors, or by means of an elaborate physical model (Bouchard 1989,Kuc and Siegel 1987). Detailed physical models have also been developed for other sensors, such as pressure transducers (EerNisse 1987). At least in the case of inductive sensors, detailed physical models lead to computation times that preclude robot simulation in interactive mode. They may be useful for calibrating more approximate models for use in robot simulation (and of course, they may be useful directly for sensor design).

Finally, it should be observed that some systems which claim *to permit* modelling of sensors do only that, that is, an interface is provided which permits the user to define sensors and their interaction to the environment. This is desirable; however, it leaves open the often difficult problem of how the simulation should actually be done.

References

Andersson, L.-E., Hurteau, G. and Stewart, N. F. A maximal distance result of interest in robotic simulation. *Applied Mathematics and Optimization* 16, 217-226, 1987.

Andersson-Stewart88 Andersson, L.-E. and Stewart, N. F. Maximal distance for robotic simulation: the convex case. *JOTA* 57, No. 2, 215-222, 1988.

André, G. Conception et modélisation de systèmes de perception proximétrique. Application à la commande en téléopération. Thèse de Docteur-Ingénieur, Rennes, France 1983.

Armstrong, W. W., Green, M. and Lake, R. Near-real-time control of human figure models. *IEEE C. G. and A.* 52-61, June 1987.

Bobrow, J. E. A direct minimization approach for obtaining the distance between convex polyhedra. *The International Journal of Robotics Research* 8, No. 3, 65-76, 1989.

Bouchard, P. Simulation et mod\'elisation de capteurs en environnement robotique. M\'emoire M.Sc., Université de Montréal 1989.

Boyse, J. W. Interference detection among solids and surfaces. *CACM* 22, No. 1, 3-9, January 1979.

Cameron, S. A. Modelling Solids in Motion. PhD thesis, University of Edinburgh, 1984.

Craig, J. J. Anatomy of an off-line programming system. *Robotics Today* 45-47, February 1985.

EerNisse, E. P. Theoretical modeling of quartz resonator pressure transducers. *41st Annual Symposium on Frequency Control*, Philadelphia, PA, 339-343, May 1987.

Fu, K. S., Gonzalez, R. C. and Lee, C. S. G. *Robotics: Control, Sensing, Vision and Intelligence*. McGraw-Hill, 1987.

Gourret, J.-P., Magnenat-Thalmann, N. and Thalmann, D. Simulation of object and human skin deformations in a grasping task. *SIGGRAPH* 21-30, 1989.

Hurteau, G. and Stewart, N. F. Collision prevention in robotic simulation. To appear, *INFOR*, 1990.

Johnson, R. H. *Solid Modeling*. North-Holland, 1986.

Joskowicz, L. Reasoning about shape and kinematic function in mechanical parts. Robotics Report No. 173, Courant Institute, New York University, September, 1988.

Kuc, R. and Siegel, M. W. Physically based simulation model for acoustic sensor robot navigation. *IEEE P. A. M. I.* Vol PAMI-9, No. 6, 766-778, November 1987.

Leven, S. and Sharir, M. Intersection and proximity problems and Voronoi diagrams, in *Advances in Robotics*, J. T. Schwartz and C.-K. Yap, eds. Lawrence Erlbaum Associates, 1987.

Luenberger, D. G. *Introduction to Linear and Nonlinear Programming*. Addison-Wesley, 1973.

Magnenat-Thalmann, N. and Thalmann, D. *Computer Animation, Theory and Practice*. Springer-Verlag, 1985.

Mahajan, R. and Walter, S. E. Computer aided automation planning: workcell design and simulation. *Robotics Engineering* 12-15, August 1986.

Major, F., Malenfant, J. and Stewart, N. F. Distance between objects represented by octrees defined in different coordinate systems. To appear, *Computers and Graphics* 1990.

Mantyla, M. *Solid Modeling*. Computer Science Press, 1988.

Moore, M. and Wilhelms, J. Collision detection and response for computer animation. *SIGGRAPH* 289-298, 1988.

Novak, B. Robotic simulation facilitates assembly line design. *Simulation* 298-299, December 1984.

Pai, D. K. and Leu, M. C. Ineffabelle--an environment for interactive computer graphic simulation of robotic applications. *IEEE Conf. on Robotics and Automation*, San Francisco, CA, 897-903, April 1986.

Paul, R. P. *Robot Manipulators: Mathematics, Programming and Control*. MIT Press, 1981.

Preparata, F. P. and Shamos, M. I. *Computational Geometry*. Springer-Verlag, 1985.

Raczkowsky, J. and Mittenbuehler, K. H. Simulation of cameras in robot applications. *IEEE C. G. and A.* 16-25, January 1989.

Red, W. E. Minimum distances for robot task simulation. *Robotica* 1, 231-238, 1983.

Samet, H. and Tamminen, M. Computing geometric properties of images represented by linear quadtrees. *IEEE P. A. M. I.* Vol PAMI-7, No. 2, 229-240, 1985.

Samet, H. *Applications of Spatial Data Structures*. Addison-Wesley, 1990.

Samet, H. *The Design and Analysis of Spatial Data Structures*. Addison-Wesley, 1990.

Schacter, B. J. Computer image generation for flight simulation. *IEEE C. G. and A.* 1, No. 4, 29-68, 1981.

Schwartz, J. T., Sharir, M. and Hopcroft, J. eds. *Planning, Geometry, and Complexity of Robot Motion*. Ablex Publishing, Norwood, New Jersey, 1987.

Segura, S. and Schrive, E. Rapport Annuel du LAMM, Universit\'e de Montpellier, France, 1985.

Sharir, M. Algorithmic Motion Planning in Robotics. *IEEE COMPUTER* 9-20, March 1989.

SIEL SV2.6 Manuel d'utilisation. INRIA, Rennes, France, January 1987.

Stauffer, R. N. Robot System Simulation. *Robotics Today* 81-90, June 1984.

Stewart, N. F. Solid Modelling. This volume, 1990.

Tilove, R. B. A null-object detection algorithm for Constructive Solid Geometry. *CACM* 27, No. 7, 684-694, July 1984.

Uchiki, T., Ohashi, T. and Tokoro, M. Collision detection in motion simulation. *Computers and Graphics* 7, No. 3 4, 285-293, 1983.

Wilhelms, J. Toward automatic motion control. *IEEE C. G. and A.* 11-22, April 1987.

Wilhelms, J. Using dynamic analysis for realistic animation of articulated bodies. *IEEE C. G. and A.* 12-27, June 1987.

Yap, C.-K. Algorithmic Motion Planning, in *Advances in Robotics*, J. T. Schwartz and C.-K. Yap, eds. Lawrence Erlbaum Associates, 1987.

18

Characteristics of a Model for the Representation of a Production Installation

Federico Sternheim
Swiss Federal Institite of Technology, Lausanne

This chapter presents the characteristics of a model for representing automated production installations. The importance of these characteristics became evident during the realization of a simulator incorporated into the chain of design of a product dedicated to automated assembly. This includes all steps from the conception of the product to the realization of the installation.

18.1. Introduction

When a designer studies a computer aided assembly installation, he needs a mathematical model capable of carrying out a virtual representation of the installation. A good model must be able to simulate, without simplifying them, the important and necessary phenomena for getting a simulation which is truly representative (REALISTIC MODEL). Another requirement is that the model allows us to simulate a particular function using elements very close to what is done in reality. This greatly simplifies the passage to the real world in the final realization (CLOSE MODEL).

In the domain of the three-dimensional simulation of cells, the first developments came following the efforts to computerize the off-line programming of robots. Packages such as "Mc Auto" (McDonnel Douglas 1982, see: Hammond et al. 1985), "Robcat" (Tecnomatix 1986), "Catia" (Dassault Système/IBM 1983), "Robosim" (General Electric Corp./Calma 1986), "Robot Programming" (Intergraph 1986) can manage complex machines, for movement control, such as robots during the tracking of trajectories defined at the level of a terminal organ (grip, tool, etc.). These are problems involving the conversion of coordinates in the direct and inverse geometric models (see Chapter 9 of this book, Denavit and Hartenberg 1955; Paul 1981; Sternheim 1987, 1988a; Reboulet). These programs can treat concurrent tasks, and the introduction of the notion of time as a law of motion allows us to extract the cycle time, useful information for evaluating the economy of operation. The use of post-processors allows us to prepare the command file submitted to the controller of the robot. What could be criticized is the orientation of these developments to the off-line programming of robots (Carter 1987; Ruokangas and Martin 1985), leaving them deficient for the domain of automated assembly.

18.2. Two simulation structures

The study of an installation of automated assembly must satisfy the demands of feasibility and economy. This study revolves around the "product" and the "installation." The aspect "product" concerns the product being assembled and the technique of assembly; a product is conceived with a view to its automated assembly. This essentially means a three dimensional geometry analysis and consideration of the basic operations and their costs. The second aspect, "installation," involves the installation which carries out the automated assembly. The notion of time, permitting the evaluation of profitability, is introduced.

These are two overlapping aspects which should be considered together. This can create difficulties because the designer of a product is often not interested early in considering the final stage, the installation, esteeming premature the consideration of the associated constraints and fearing the loss of all initiative. But, it must be possible to fulfill the automatic assembly of the product in question, while respecting the imperatives of feasibility and economy. What often happens is that a planned product is later adapted to automated assembly. There is a very little difference between the adaptation of already existing products and the adaptation of those designed without taking automation into account. Briefly, if the "installation" aspect is neglected, problems of profitability may occur: if the "product" aspect is neglected, problems of feasibility may appear.

In the next sections, we discuss the importance of the quality of the internal mathematical model, in the context of a study for the design of a CAD assembly installation.

18.2.1. Utility of a simulation

Once the assembly procedure is defined and the feasibility of all the basic operations proved, two approaches permit a preliminary study of a complex automated installation.

The first approach consists in emphasizing the flow of the production and using the most simple automated stations to guarantee a high reliability of each unit. Techniques of operation research will be available. A simulation will confirm if the predictions are correct and will also permit the aspects that operation research cannot take into account.

In the second approach, the layout is designed and the behavior of the cell simulated using the experience of the designer and the documentation of the suppliers. This may seem a simplification of the first approach, but the process is slightly different. This approach can be very efficient if the designer has enough experience which he can translate into simple rules.

Currently, the studies of cell design using the first approach are rare and incomplete. The second approach is beginning to be used in applications targeting the off-line programming of manipulators and robots. Anyhow, whatever the approach, simulation represents the final verification before starting a concrete realization. It is necessary to have, most especially, a modelling tool able to simulate realistically and in detail the behavior of the cell.

18.2.2. A model for an assembly installation

The off-line component of a model for automated assembly does not need to be highly evolved. But, it presents particularities that must be added to the previous model. We only present here those which correspond to the installation without going into the aspect of the product itself. The characteristics that we require from a cell simulator in the domain of off-line programming of robots are the following ones:

(1) The environment must be three dimensional. Simplifying approximations hinder too greatly the understanding of the behavior of the simulated system and the rapid setting up of the installation.

(2) The simulator must combine the notion of time and the law of motion (fourth dimension). This is the only way of controlling the global execution of the task and the synchronization of the basic operations. Only the notion of a law of motion permits a detailed study. Also, these properties allow us to get the cycle time with precision.

(3) A three dimensional graphic representation should be available to present a synthesis of the behavior of the installation during the simulation. This feature has the advantage of communicating very quickly an enormous volume of information. This does not mean making a visual representation. The primary goal of the simulation is to study, by the intermediary of a mathematical model, the behavior of the installation.

(4) The simulator must be extensible without major changes. Primitives and facilities must be supplied which simplify the integration. In this domain, all modelling which is not three dimensional is not extensible (Wesley et al. 1980)

(5) A single simulation language should be available for describing the execution of the operations and managing objects and machines with heterogeneous functions and characteristics. The simulator should not need a lengthy period of apprenticeship of the languages used to control the machines under consideration. The efficiency of the simulator and the ease of adaptation to the real world in the final phase will depend on the simulation language.

(6) For the coordinate transformations of the robots, the algorithms of the inverse and direct geometric models of the CAD tool must behave in complete analogy with the real controller. In off-line programming and in most tasks carried out with the help of robots, not respecting this rule makes the CAD tool useless for the preparation of the task (Denavit and Hartenberg 1955; Castelain et al. 1986; Dillman et al. 1986; Dombre and Fournier; Sternheim 1988b).

(7) It must be possible to sketch and simulate how the kinematics works to better define the specifications of a new component. The detailed development of a new robot, for example, must be eased by the environment available in the utility (Sternheim 1987, 1988a).

We must add to these basic characteristics the requirements imposed by the installation of the automated assembly. The following list comes from experience acquired during the development of a project "Spirit." A good internal model for automated assembly should meet the following criteria:

(8) It must be highly able to process many concurrent tasks and moving objects. Each object (component, palette, machine, transport system, etc.) must be manageable without limiting the number or the level of the interactions. It is necessary to emulate the complete installation and not single machines (Sternheim 1989).

(9) It must be able to model complex systems of conveyors and various systems of storage to reproduce faithfully and easily the flow of circulating parts. This is the most difficult aspect to implement in the CAD tool but it is also the most useful for the setting up of the installation (REALISTIC MODEL) (Sternheim 1989).

(10) It must allow, using the familiar elements (CLOSE MODEL), the circulation of a large flow of information. On the quality of these elements (for example, the way of handling an interruption or the way of connecting two machines, etc.) will depend the ease of adapting to the real world when the installation is realized. We believe they should be as close as possible to reality.

(11) It must provide an intelligent and coherent connection with the "product" side of the simulation, which will handle assembly planning and the design of the junctions for the automated assembly. For the reasons mentioned in the introduction, this connection will be directional. The dialogue between the aspects "product" and "installation" will depend upon the quality of this link.

(12) It must be able to consider the geometric uncertainties which appear in the kinematic chain during an assembly operation. This covers errors resulting from variations of form (tolerance, temperature, etc.) and errors because of the play, and involves having an efficient way of testing whether the connections between the components of the product have been correctly dimensioned. This is an important step to ensure the feasibility of the automated assembly, often with repercussions on the resources used.

(13) During the simulation of the functioning of the installation, the reduction of the performance of the computer should be proportional to the increasing of the complexity of the installation. A too fast reduction indicates a bad model especially as far as the management of the transport and storage of pieces.

There is an important difference between a simulation of the off-line programming of robots and a simulation for the study of an assembly installation: in the former each sub-component of the problem can be studied separately, while in the latter it is important to treat the problem globally. Moreover, in the domain of assembly, the number of objects involved is significantly greater. Global treatment of an installation involving heterogeneous resources also brings into play the command strategy and the manipulation of the resulting flow of information. The quality of the simulation language, used to describe the behavior of the installation, plays an important role in the study of the distribution of intelligence in the installation (local intelligence, distributed intelligence, hybrid system). A variety of command architecture must be studied and all changes must be feasible very fast. While it is true there is no comprehensive theory which provides the specification of the command hierarchy, a simulator can, even so, give good results for installations from simple to average complexity. For the very complex cases, it would be necessary to decompose the problem into modular blocks and impose conditions at the limits.

There is also a class of phenomena that we should be able to study. Problems tied to the starting up of an installation (cold start-up), starting up after an unforeseen interruption (warm start-up) and other transitory regimes merit attention, with the possibility of introducing different assembly lines which run concurrently. Lastly, the user should be made aware of the risks of destroying important variables in the flow of information circulating in a cell. This can affect production and the security of the installation.

Clearly the need to control several objects and a quantity of information largely superior to the off-line case has repercussions on the power of the computing hardware for computation and graphics performance. It is this factor which will largely determine the size and complexity of the installation that we will be able to simulate (see Figures 18.1-2-3). The continuous evolution of computer performances and the coming of parallel architectures for graphics workstations permit us to stay optimistic about the problem.

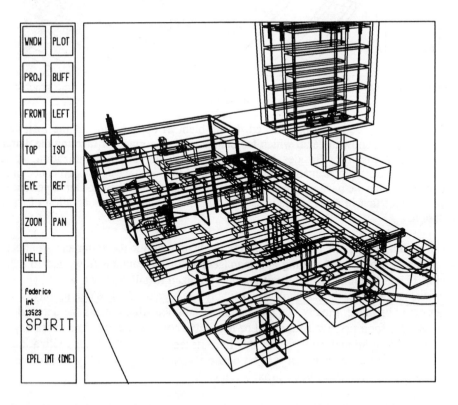

Figure 18.1. Example of simulation involving 53 concurrent tasks, 27 "machine" entities, 164 sensors: nearly 100 objects moving in a four-dimensional environment.

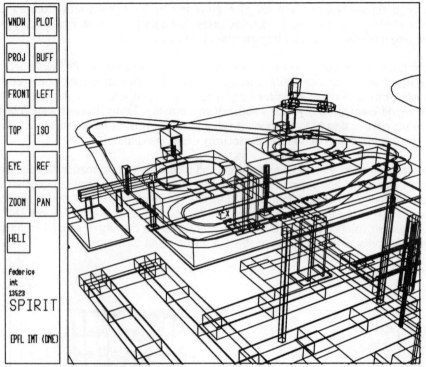

Figure 18.2. Object representation is extremely simplified to allow real-time graphics animation

In this chapter, we have not attacked the problem of optimization of the installation, because we do not know in detail which factors are truly important for meeting the demands of economy and feasibility. Besides, before trying to find methods of optimization, we must anyhow define a truly representative model. The model that we have developed is not an ideal one in the sense that we have ignored the effects of friction and ballistics. Our experience has shown, however that in most studied cases, it furnishes a useful representation of the real installation.

With such a model, it is possible to design the layout, define the command architecture, solve different problems of coding and management of information flow, and to test the behavior. If it is true that we shall not get an optimized installation, we will however, be able to evaluate whether the threshold of profitably has been reached. When problems occur, changes could be made. Finally, we will get an installation which meets the demands, and the time needed for a first evaluation will be negligible compared to the time necessary for a study using conventional methods or the time necessary for the final realization. No preliminary purchases of material will be necessary, and the probability of accident and damage during the first tests will be strongly reduced.

It is desirable that the model, capable of treating a great variety of aspects, from the synchronization of concurrent tasks to the management of transport systems, from the study of the command architecture to the tests of the coding techniques for the palettes, be unique.

This simplifies the integration and makes the design of new installations easier. Furthermore, it should not contain approximations which move it away from reality, even at the cost of increasing its complexity. A typical example of such an approximation is found in the various utilities on PC's for the simulation of transport systems (AGV's and palettes) which manage these components as shifts registers. Such an approximation permits a spectacular simplification of the internal model thanks to the elimination of the notion of law of motion, which in turn makes the three dimensional environment superfluous. Under these conditions, the results of the simulation are useless for an in-depth study and it would be utopian to speak of optimization.

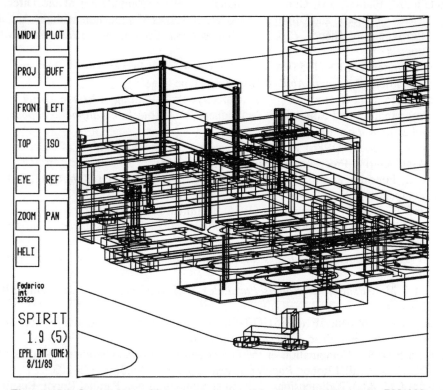

Figure 18.3. It took about 30 hours to prepare this virtual model. Apollo DN4000 and DN4500 workstations were used for both modeling and animation. Apollo DN10000 Prism supercomputers are needed to perform real-time animation during simulation in this case study.

It is not until a high quality model is defined that it seems reasonable to seek to optimize it with methods of operations research.

18.3. Conclusion

It is difficult to speak of optimization when the parameters involved are unknown or not understood in detail. First of all, a good model is necessary. This makes the difference between an inefficient tool and one which is truly useful for design. Then, if possible, the introduction of an optimization technique will allow us to switch from a tool which supplies aid at conception to a true design tool. Today, we have a useful model for assemblage installations, and now, it will be possible to accelerate the design of a profitable, even if not perfectly optimized installation.

References

Carter S (1987) OFF-LINE Robot Programming: The State-of-the-Art, *The Industrial Robot*, Vol.14, No4, pp.213-215.

Castelain JM, Flamme JM, Gorla B, Renaud M (1986) Computation of the Direct and Inverse Geometric and Differential Models of Robots Manipulators with the Aid of the Hypercomplex Theory, *Proc. 16th International Symposium on Industrial Robots*, Brussels

Denavit J, Hartenberg RS (1955) A Kinematic Notation for Lower-pair Mechanisms Based on Matrices, *Journal of Applied Mechanics*, June Issue

Dillman R, Hornung R, Huck M (1986) Interactive Programming of Robots Using Textual Programming and Simulation Techniques, *Proc. 16th International Symposium on Industrial Robots*, Brussels

Dillman R, Huck M (1986) A Software System for the Simulation of Robots Based Manufacturing Processes, *Robotics*, Vol.2, pp.3-18.

Dombre E, Fournier A *Conception Assistée par Ordinateur en Robotique*, LAMM, Montpellier

Hammond R, Hansen J, Hansen L (1985) Elimination of Application Risk Utilizing MCAUTO PLACE Software and Corporate Robot Facilities, *Proc. Conf Robot 9* Detroit

Paul RB (1981) *Robots Manipulators: Mathematics, Programming, and Control*, The MIT Press.

Reboulet C *Modélisation de Robots Parallèles*, CERT/DERA, 2 av. Ed. Belin, 310555 Toulouse

Ruokangas CC, Martin JF (1985) Off-line Programming: A Robotics Space Shuttle Welding System, *Proc. Conf. Robot 9*, Detroit.

Sternheim F (1987) Computation of the Direct and the Inverse Geometric Models of the DELTA4 Parallel Robot, *Robotersysteme*, Springer, Vol.3, pp.199-203.

Sternheim F (1988a) Tridimensional Computer Simulation of a Parallel Robot, Results for the DELTA4 Machine, *Proc. 18th International Symposium on Industrial Robots*, Lausanne, October 1988.

Sternheim F (1988b) Simulation 3D sur écran, *Proc. 18th International Symposium on Industrial Robots*, Lausanne.

Sternheim F (1989) Dispatching and Transit Simulation (DTS) in a 4D-Environment, *Proc. 20th International Symposium on Industrial Robots*, Tokyo.

Wesley MA, Lozano.Perez, Liebermann LI, Lavin MA, Grossman DD (1980) A Geometric Modeling System for Automated Assembly, *IBM Journal Research and Development*, Vol.24, No1.